"Required reading for every psychologist. With the help of William Stern's critical personalism, James Lamiell clearly explains the devasting methodological error at the heart of contemporary psychology and the way out of the quagmire in order to better understand persons and pressing social problems."

Brian Schiff, *Esmond Nissim Professor of Psychology at The American University of Paris, France*

Primer in Critical Personalism

This insightful book offers contemporary psychologists and other social theorists an understanding of the comprehensive system of thought developed by the German scholar William Stern (1871–1938) known as critical personalism.

Expanding the author's ongoing efforts in this area, the book considers, firstly, how critical personalism could ground a needed revival of psychological science, a need created by the field's gradual transformation, through its widespread adoption of aggregate statistical methods of investigation, into a discipline better characterized as 'psycho-demography.' Consistent with Stern's own view of the potential of critical personalism *vis-a-vis* socio-ethical concerns, the book then explores how the framework could facilitate a transcendence of thinking about racial and other social relationships beyond currently prevailing narratives about *personkinds* into narratives that are actually about *persons.* This part of the book includes a chapter discussing Stern's own historical efforts in this direction, serving to highlight the non-individualistic nature of critically personalistic thinking. Throughout, Lamiell constructs a clear case for the merits and applicability of critical personalism in modern psychology and social thought.

Primer in Critical Personalism will interest established psychological scientists and advanced students in the field, as well as those who are concerned about our contemporary socio-cultural ethos and the prospects for its improvement, including philosophers, sociologists, educators, journalists, clerics, and thoughtful laypersons alike.

James T. Lamiell is a Professor Emeritus in the Psychology Department at Georgetown University, USA. His scholarly interests are in the history and philosophy of psychology, the psychology of subjective personality judgments, and methodological issues pertaining to psychological research.

Advances in Theoretical and Philosophical Psychology
Series Editor: Brent D. Slife

Hermeneutic Approaches to Interpretive Research: Dissertations In a Different Key
Philip Cushman

A Humane Vision of Clinical Psychology, Volume 1: The Theoretical Basis for a Compassionate Psychotherapy
Robert A. Graceffo

A Humane Vision of Clinical Psychology, Volume 2: Explorations into the Practice of Compassionate Psychotherapy
Robert A. Graceffo

A Philosophical Perspective on Folk Moral Objectivism
Thomas Pölzler

A Psychological Perspective on Folk Moral Objectivism
Jennifer Cole Wright

Suffering and Psychology
Frank C. Richardson

Posttraumatic Joy: A Seminar on Nietzsche's Tragicomic Philosophy of Life
Matthew Clemente, Edited with Introduction by Andrew J. Zeppa

Towards the Psychological Humanities: A Modest Manifesto for the Future of Psychology
Mark Freeman

Primer in Critical Personalism: A Framework for Reviving Psychological Inquiry and for Grounding a Socio-Cultural Ethos
James T. Lamiell

For more information about this series, please visit www.routledge.com/Advances-in-Theoretical-and-Philosophical-Psychology/book-series/TPP

Primer in Critical Personalism

A Framework for Reviving Psychological Inquiry and for Grounding a Socio-Cultural Ethos

James T. Lamiell

Routledge
Taylor & Francis Group

NEW YORK AND LONDON

First published 2024
by Routledge
605 Third Avenue, New York, NY 10158

and by Routledge
4 Park Square, Milton Park, Abingdon, Oxon OX14 4RN

Routledge is an imprint of the Taylor & Francis Group, an informa business

British Library Cataloguing-in-Publication Data
A catalogue record for this book is available from the British Library

Library of Congress Cataloguing-in-Publication Data
Names: Lamiell, James T., author.
Title: Primer in critical personalism : a framework for reviving psychological inquiry and for grounding a socio-cultural ethos / James T. Lamiell.
Description: Abingdon, Oxon ; New York, NY : Routledge, 2024. | Series: Advances in theoretical and philosophical psychology | Includes bibliographical references and index.
Identifiers: LCCN 2023052319 (print) | LCCN 2023052320 (ebook) | ISBN 9781032450551 (hardback) | ISBN 9781032450568 (paperback) | ISBN 9781003375166 (ebook)
Subjects: LCSH: Personality. | Personalism--Psychological aspects.
Classification: LCC BF698 .L37158 2024 (print) | LCC BF698 (ebook) | DDC 155.2--dc23/eng/20231218
LC record available at https://lccn.loc.gov/2023052319
LC ebook record available at https://lccn.loc.gov/2023052320

ISBN: 978-1-032-45055-1 (hbk)
ISBN: 978-1-032-45056-8 (pbk)
ISBN: 978-1-003-37516-6 (ebk)

DOI: 10.4324/9781003375166

Typeset in Times New Roman
by MPS Limited, Dehradun

This book is dedicated to the memory of William Stern (1871–1938), and to the efforts of those scholars who, succeeding him, have worked in his intellectual patrimony, attempting to further critical personalism as a framework for psychological science and as a foundation for understanding human social interaction.

Contents

Acknowledgments   *xiii*

Advances in Theoretical and Philosophical Psychology   *xiv*

Preface   *xvi*

PART I

Rudimentary Considerations   1

1   **On the Need to Revive Psychological Inquiry: The Historic Transformation of Empirical Psychology into Psycho-Demography**   3

Introduction 3

On the Original Structure of Experimental Psychology 6

An Illustration of Early Experimental Psychology: Basic Experiments in Mental Chronometry 7

The Beginnings of Psychology's Transformation into Psycho-Demography 9

The Establishment of 'Differential' Psychology 9

Experimental Psychology's Gradual Transformation into a Species of Differential Psychology 11

*Understanding Why Knowledge about Individual
Differences in Psychological Doings Is Not
Knowledge about the Psychological Doings of
Individuals 14*

2   Critical Personalism: Its Core Philosophical Tenets   19
*An Important Cautionary Note: Personalism
Is Not Individualism 20
The Concept of the Person: Rudimentary
Considerations 20
    What Is a Person? On the Nature of Personal
        Being from a Critically Personalistic
        Perspective 22
    A Person Is a Unitary, Psychophysically
        Neutral Being 23
    A Person Is a Causally Effective Agent 24
    A Person Is an Inherently Evaluative Being 26
    A Person Is a Distinctive Individuality 26
    Who Is This Person? On the Basic Dynamics of
        Psychosocial Development from a Critically
        Personalistic Perspective 28*

**3   The Challenge of Reviving Psychological Studies:
Some Further Historical Perspectives and Some
Possibilities for Moving Forward**   33
*From Psychological Science to a Scientistic
    'Psychology' 35
    A Confusion within a Confusion 40
Reviving Psychological Science 42
    Restructuring Psychological
        Experimentation 43
    Expanding the Space for Qualitative
        Investigations 45
    Reviving a Broad Conception of 'Science'
        as 'the Making of Knowledge' 47*

PART II
Toward a Critical Inter-personalism in the
Grounding of a Socio-Cultural Ethos   51

4   Echoes of William Stern's Socio-Cultural Voice   53
On the Ethical Significance of Tolerance 53
Echoes of Stern's Socio-Cultural Voice
   in the Domain of Child Psychology 55
      On the Development of Lying in Children 56
      On the Identification of Highly Talented
         Pupils 58
      On the Practice of Psychoanalytic
         Psychotherapy with Children and
         Adolescents 59
Echoes of Stern's Socio-Cultural Voice in the
   Domain of Psychological Testing 62
      On Psychological Testing as a Socio-Cultural
         Issue 62
      Socio-Cultural Concerns about the
         Proliferation of Psychological Testing 65
Conclusion 67

5   Some Critically Personalistic Observations on
Current Discussions of Racism in American
Society   71
The Impersonal Nature of Discourse about
   Personkinds 71
      Some General Considerations 71
      A Brief Sojourn into Some Relevant History 73
      Statism in White Fragility 77
Other Impersonal Aspects of Contemporary
   Discourse about Race in America 80
      Mechanistic Narratives Concerning the Social
         Dynamics of Systemic Racism 80

Impersonal Understandings of the Concept of
Personal Identity 82
On the Possibility of Colorblindness in
Contemporary Social Exchanges 85

6   **Toward a Broadened Perspective: Navigating**
     Some Conceptual Obstacles to Critically
     Personalistic Thinking   90
Full Disclosure: Confession of a Concrete
Idealist 90
Some Prominent Obstacles to Critical
Personalism—Then and Now 91
Early 20th-Century Behaviorism 91
Contemporary Statism 92
Other Troublesome Aspects of Statist
Thinking 93
Prevailing Misunderstandings of
Probability 94
Concern with Typicality 94
Agenting of Personkinds 95
Empiricist Understandings of the Saying
'Everyone Is Different' 97
A Rationalist Understanding of Societal
Distinctiveness 99
Respecting the Diversity of Personkinds While
Recognizing the Overarching Commonality
of Humans 100

Index   *106*

Acknowledgments

I am very thankful to my co-participants in a bi-racial discussion group that was formed in June 2020 to discuss the enduring problem of anti-Black racism in America. Their sustained commitment, both individually and collectively, to sharing experiences of and perspectives on racism has strengthened my resolve to write about the way in which critically personalistic thinking could improve the manner in which people engage with one another regarding racial issues and other matters of consequence for social life. Our discussions had particular influence on my decisions about what topics to single out for consideration in Chapter 5.

Of course, I am also deeply grateful to my wife, Leslie, for her patience and understanding during the many months of my preoccupation with this project. It is difficult for me to see how I could have completed the work without the time and space that her consideration afforded me.

Advances in Theoretical and Philosophical Psychology

Series Foreword

Brent D. Slife Series Editor

Psychologists need to face the facts. Their commitment to empiricism for answering disciplinary questions does not prevent pivotal questions from arising that cannot be evaluated exclusively through empirical methods, hence the title of this series: *Advances in Theoretical and Philosophical Psychology*. For example, such moral questions as, 'What is the nature of a good life?' are crucial to psychotherapists but are not answerable through empirical methods alone. And what of these methods? Many have worried that our current psychological means of investigation are not adequate for fully understanding the person (e.g., Gantt & Williams, 2018; Schiff, 2019). How do we address this concern through empirical methods without running headlong into the dilemma of methods investigating themselves? Such questions are in some sense philosophical, to be sure, but the discipline of psychology cannot advance even its own empirical agenda without addressing questions like these in defensible ways.

How then should the discipline of psychology deal with such distinctly theoretical and philosophical questions? We could leave the answers exclusively to professional philosophers, but this option would mean that the conceptual foundations of the discipline, including the conceptual framework of empiricism itself, are left to scholars who are *outside* the discipline. As undoubtedly helpful as philosophers are and will be, this situation would mean that the people doing the actual psychological work, psychologists themselves, are divorced from the people who formulate and re-formulate the conceptual foundations of that work. This division of labor would not seem to serve the long-term viability of the discipline.

Instead, the founders of psychology—scholars such as Wundt, Freud, and James—recognized the importance of psychologists in formulating their own foundations. These parents of psychology not only did their own theorizing, in cooperation with many other disciplines; they also realized the significance of psychologists continuously *re*-examining these theories and philosophies. This re-examination process allowed for the people most directly involved in and knowledgeable about the discipline to be the ones to decide *what* changes were needed, and *how* such changes would best be implemented. This book series is dedicated to that task, the examining and re-examining of psychology's foundations.

References

Gantt, E., & Williams, R. (2018). *On hijacking science: Exploring the nature and consequences of overreach in psychology.* London: Routledge.

Schiff, B. (2019). *Situating qualitative methods in psychological science.* London: Routledge.

Preface

Over the course of an estimable scholarly career harshly ended by Nazism, and then overlooked by most of posterity, the German philosopher and psychologist William Stern (1871–1938) made many and highly varied contributions to the project of the human sciences generally, and to psychology in particular. By his own account, the centerpiece of those contributions was a comprehensive system of thought—a *Weltanschauung* or 'worldview'—that he called *critical personalism*. In the concluding sentence of a book published in its first edition in 1918, titled (in translation) *The Human Personality,* Stern made explicit his firm conviction that critical personalism offered a 'conception of human nature [that would prove] fruitful both for work in the humanities and for the grounding of cultural life' (Stern, 1923, p. 270, brackets added).[1]

During my own scholarly career, I, too, have come to believe strongly in the power of critically personalistic thought as a framework for conducting and evaluating psychological studies, as a guide for understanding selves, and as a foundation for social interactions and community life. The present volume issues from this conviction, and furthers my efforts over many years (e.g., Lamiell, 2003; 2010a; 2010b; 2020; 2021), to broaden and deepen familiarity with Stern's views among other contemporary thinkers—philosophers, theologians, social theorists, journalists, and reflective laypersons alike.

To be sure, my own efforts thus far in this direction have been complemented by works of other contemporary scholars among the relatively few worldwide who have first-hand knowledge of Stern's writings. Within just the past three decades, there have appeared significant works authored or edited by Bühring (1996), Deutsch (1991), Heinemann (2016), Koczanowicz-Dehnel (2017), and Lück and Löwisch (1994). However, one of those works was published in Polish (Koczanowicz-Dehnel, 2017), and all of the others in German— languages far less accessible worldwide than English. An additional hindrance to the spread of conversance with critical personalism may well have been the way in which Stern's thinking was characterized in a widely read survey of personalistic thought authored by the theologian Rufus Burrow, Jr. (birth year unknown; death in 2021) and published in English in 1999 under the title *Personalism: A Critical Introduction* (Burrow, 1999).

Without doubt, Burrow's book has much to recommend it to readers interested to learn more about the history and current status of personalistic thought, and if the present volume prompts readers who have not heretofore done so to look into Burrow's work, so much the better. However, out of some 256 pages of text (apart from notes and bibliographic material), Burrow devoted a scant one and a half pages to a discussion of Stern's ideas. In those pages, which appeared in a chapter titled 'Some Less Typical Types of Personalism,' Burrow marginalized the relevance of Stern's ideas to what he (Burrow) regarded as mainstream personalistic thinking, and, at that, he failed to accurately characterize Stern's ideas in their own right (cf. Burrow, 1999, pp. 38–40). Subsequently, in what was possibly a ripple effect of Burrow's dismissal of Stern's views, no mention of them at all is to be found in a book authored more recently by the Spanish philosopher Juan Manuel Burgos (b. 1961). That work, titled *An Introduction to Personalism,* first appeared in Spanish in 2012 and then in an English translation by R. T. Allen in 2018 (Burgos, 2018). Burgos's failure to include any discussion of critical personalism is all the more surprising given that he devoted some 20 pages of his book to 'personalism in the German language' (Burgos, 2018, p. 119). There he included a brief sketch of the work of Edith Stein (1891–1942), who was directly, if only briefly, exposed to Stern's tuition through coursework she undertook during her pre-doctoral studies at the University of Breslau (now Wroclaw, Poland). Stein went on to pursue her doctoral studies in philosophy under Edmund Husserl (1859–1938).[2]

All of these considerations point to the need for continued efforts to acquaint contemporary thinkers with the central tenets of Stern's critical personalism. Part I of the present volume has been written with this as the primary objective. Chapter 2 addresses two broad questions: What is *a* person? And who is *this* person? There, I have sought to provide readers with an accurate but accessible presentation of the critically personalistic answers to those questions. Chapter 3 is concerned with the challenge of reviving psychological inquiry within the framework of critical personalism. There, two major questions are taken up: (1) What are the basic knowledge objectives of psychological studies? (2) What sort of inquiry can legitimately be said to conform to those knowledge objectives?

For at least some readers of this volume, my mention of 'reviving' psychological studies will raise eyebrows. Why should I write of 'reviving' psychological inquiry? That is: why would it now, well into the 21st century, appear necessary to clarify the knowledge objectives and apposite investigative methods of a field of inquiry first prosecuted as an empirically oriented discipline in the latter part of the 19th century? It is just this question to which Chapter 1 is addressed. That chapter offers a condensed version of an argument, developed at length in my 2019 book (Lamiell, 2019), that over the course of the 20th-century mainstream psychology gradually abandoned investigative methods logically suited to the advancement of our understanding of individuals' psychological 'doings'—their sensations, perceptions, judgments, emotions, cognitions, memories, attitudes, behaviors, etc.—the original knowledge objective of systematic psychological research, in favor of aggregate, statistical methods of inquiry that, however useful they may be for the pursuit of other knowledge objectives, are fundamentally and irremediably ill-suited to gaining knowledge of individuals. In effect, I argue, most of the discipline now referred to as 'psychology' is actually no longer a genuine *psychology* at all, but has become instead, *de facto,* a species of demography that I have branded 'psycho-demography.' Once the real epistemic consequences of this paradigm shift within psychology are grasped, the need to *revive* psychological studies becomes obvious. While critical personalism is not the only framework within which such a revival could be undertaken, it is certainly one such framework, and, at that, one highly deserving of the serious consideration that has for so long eluded it.

Beyond all of the foregoing, a second major purpose of this work is to probe further Stern's profound belief, expressed in the quotation cited in the opening paragraph of this work, that critically personalistic thinking could serve as a 'grounding for cultural life.' This has been my mission in Part II of this volume. Chapter 4 offers a historical look at several different contexts in which William Stern himself lent his critically personalistic voice to discussions of socio-cultural issues that arose during his time.

The substantive concern in Chapter 5 is with what is arguably the most pressing social issue in American society today: racism. Currently, discourse on this topic, both in the writings of social scientists and in less formal works aimed at popular audiences, is dominated by distinctly *im*personal narratives, and thus provides an excellent medium for contrasting such narratives with those that would issue from a critically personalistic perspective.

Chapter 6, the book's last, aims to broaden the perspective beyond racial matters, emphasizing ideas that would guide critically personalistic discourse more generally and thus ground socio-cultural life in the way that I believe Stern envisioned.

Naturally, this book has deep roots in the teaching and research experiences that I gained over the course of a 42-year career as a university professor. Most of my pedagogical and scholarly efforts during that time were focused on (a) methodological/epistemological issues in psychological research generally, and (b) the life and works of William Stern. Indeed, my two most recent books reflect these continuing interests directly. The first of those two is the 2019 book mentioned above, which bears the title *Psychology's Misuse of Statistics and Persistent Dismissal of its Critics.* The more recent of the two appeared in 2021 under the title *Uncovering Critical Personalism: Readings from William Stern's Contributions to Scientific Psychology* (Lamiell, 2021). That work is composed primarily of my translations of selected original works by Stern that were published in German between 1906 and 1933. The book was intended to give readers (quasi) first-hand access to some of Stern's most important writings.[3]

Beyond my academic experiences, significant intellectual impetus for this particular project has come from experiences I have been having subsequent to my official retirement in 2017. Shortly after the murder of Mr. George Floyd in May of 2020, I joined with nine other individuals, five of them Black, the other four in addition to myself White, to form a bi-racial group intent on discussing in earnest the problem of race relations in American society. This group has been meeting regularly over these past three-plus years (by video technology when the Arizona summer heat and/or COVID concerns have dictated), and has continued to meet as this book was being written.

Within the context of this group's work, I and the other members have been reading or auditing/viewing a great deal of material on the history and current status of race relations in the United States, and we have been engaged in, literally, hundreds of hours of conversation about that material and a wide variety of race-related matters— including many personal experiences of group members. As we have been doing so, I have been struck repeatedly by the extent to which current discussions of the social dynamics of racism, both personal and systemic, are dominated by fundamentally impersonal narratives. I will argue that while such narratives have a rightful place, they are often unhelpful, and can even be obstructive *vis-a-vis* efforts to mitigate racism in American society moving forward. This observation has in turn prompted me to reflect upon how significantly those efforts might be enhanced by the adoption of a critically personalistic perspective.

Especially noteworthy to me in this regard is the regularity with which I have encountered expressions of the belief that the mitigation of racism would be greatly facilitated if engaged citizens would look beyond racial categories and to 'really get to know' other individuals *as persons.*

Yet, despite what appears to be broad consensus on this point in the abstract, close and careful attention to the question of just what 'getting to know other individuals as persons' would entail has been lacking. Instead, conversations have repeatedly reverted to narratives in which speakers refer both to themselves and to other individuals simply as instances of the respective races that those individuals are taken to instantiate. Discussion then proceeds largely in terms of statements about what members of different races 'typically' or 'in general' experience, think, believe, feel, do, want, etc. In this way, narratives about persons give way to narratives about *personkinds* (e.g., Blacks, Whites, males, females, etc.), yet gain credence as discourse about each of the abstract 'those' who may be seen to instantiate the designated *personkinds*. Curiously, the credibility of such narratives as informed contributions to the project of improving relationships between persons actually seems to be enhanced among discussants by the ubiquitous and deceptively modest disclaimer that 'of course, there will always be individual exceptions' to the patterns that are said to 'typically' hold for particular *personkinds*.

For reasons the reader will find clarified in this book, such disclaimers are, practically speaking, utterly empty, as is all psycho-demographic discourse in the context of efforts to advance genuinely inter*personal* understanding. But just because this crucially important point is so poorly understood, both within the general public and among professional scholars and writers on whom that general public relies for guidance, the vast and unbridgeable logical space that in fact exists between knowledge about *personkinds*, on the one hand, and knowledge about persons, on the other, is obscured. Situated squarely within that vast conceptual space lie profoundly important questions concerning our implicit assumptions about the nature of personhood itself, and about the social development of persons, i.e., about how individuals become the individually distinct persons they are at any given point in time. As long as that conceptual space remains dark—and psycho-demographic narratives cannot and will not ever illuminate it—the difference between getting to know persons, on the one hand, and characterizing individuals on the basis of what is 'typical' or 'generally true' among individuals of 'their kind,' on the other, will remain obscured. Declarations of the latter sort will continue to be mistaken for useful contributions to the realization of the former objective—the dutifully acknowledged existence of 'exceptions' to the 'rules' putatively revealed by population-level patterns notwithstanding—and the untoward consequences of this mistake for the socio-cultural fabric of communities will persist. Indeed, those consequences will often be exacerbated.

I see here the need for a work that addresses directly (a) what it would mean to 'get to know individuals as persons' from a critically personalistic perspective, (b) why aggregate, i.e., population-level, psycho-demographic thinking is not only inapt but at times obstructive *vis-a-vis* this objective, and (c) how critically personalistic thinking could provide a corrective for this and, in the process, a sound conceptual basis for a critically *inter*-personal socio-cultural ethos. The present book is the result of my effort to address these needs.

Notes

1 Here and in all subsequent quotations of original German texts throughout this volume, the translations are my own unless otherwise indicated.
2 Husserl and Stern were personally known to one another, and Husserl would eventually mentor the doctoral studies of Stern's son, Günther Anders (1902–1992).
3 To the reader of the present volume who would look into my 2021 book, the basis for my dissatisfaction with the characterization of Stern's critical personalism by Burrow (1999), as noted above, will become obvious.

References

Bühring, G. (1996). *William Stern oder Streben nach Einheit* [William Stern or the quest for unity]. Frankfurt am Main: Verlag Peter Lang.

Burgos, J. M. (2018). *An introduction to personality*, R. T. Allen (Transl.). Washington, D.C.: Catholic University of America Press.

Burrow, R., Jr. (1999). *Personalism: A critical introduction*. Nashville, TN: Chalice Press.

Deutsch, W. (Hrg.) (1991). *Die verborgene Aktualität von William Stern* [The hidden topicality of William Stern]. Frankfurt am Main: Verlag Peter Lang.

Heinemann, R. (2016). *Das Kind als Person: William Stern als Wegbereiter der Kinder- und Jugendforschung 1900 bis 1933* [The child as person: William Stern as pioneer of child and adolescent studies 1900 to 1933.]. Bad Heilbrunn, Germany: Verlag Julius Klinkhardt.

Koczanowicz-Dehnel, I. (2017). *William Stern w perspektywie nowej historii psychologii* [William Stern in the perspective of the new history of psychology]. Warsaw, Poland: Wydawnictwo Naukowe Scholar.

Lamiell, J. T. (2003). *Beyond individual and group differences: Human individuality, scientific psychology, and William Stern's critical personalism*. Thousand Oaks, CA: Sage Publications.

Lamiell, J. T. (2010a). Why was there no place for personalistic thinking in 20th century psychology? *New Ideas in Psychology*, *28*, 135–142. 10.1016/j.newideapsych.2009.02.002

Lamiell, J. T. (2010b). *William Stern (1871–1938). A brief introduction to his life and works*. Lengerich, Germany: Pabst Science Publishers.

Lamiell, J. T. (2019). *Psychology's misuse of statistics and persistent dismissal of its critics*. London: Palgrave-Macmillan.

Lamiell, J. T. (2020). William Stern (1871–1938), Eclipsed star of early 20th century psychology. In *Oxford Research Encyclopedia of Psychology*. Oxford University Press. Published online March 2020. 10.1093/acrefore/978019023655 7.013.523

Lamiell, J. T. (2021). *Uncovering critical personalism: Readings from William Stern's contributions to scientific psychology*. London: Palgrave-Macmillan.

Lück, H. E., & Löwisch, D.-J. (Eds.). (1994). *Der Briefwechsel zwischen William Stern und Jonas Cohn: Dokumente einer Freundschaft zwischen zwei Wissenschaftlern* (Correspondence between William Stern and Jonas Cohn: Documents of a friendship of two scientists). Frankfurt am Main: Verlag Peter Lang.

Stern, W. (1923). *Person und Sache: System der philosophischen Weltanschauung, zweiter Band: Die menschliche Persönlichkeit, dritte unveränderte Auflage* [Person and thing: A systematic philosophical worldview, Volume 2: The human personality, third unaltered edition]. Leipzig: Barth. [First edition 2018.]

Part I

Rudimentary Considerations

1 On the Need to Revive Psychological Inquiry

The Historic Transformation of Empirical Psychology into Psycho-Demography

Introduction

A major concern of this book lies in the domain of social relationships, with particular attention being devoted to the matter of interracial relationships in contemporary American society. From the very outset, however, it is vital to appreciate that all 'social' relationships are, at their core, inter-*personal* relationships. Consequently, an understanding of those relationships must appeal, implicitly even if not explicitly, to a basic conception of *persons*.

Psychology has long been regarded, by the field's insiders and outsiders alike, as that discipline aimed at providing knowledge about the nature of persons' psychological 'doings'—their perceptions, emotions, judgments, cognitions, memories, attitudes, behaviors, etc. The narratives found in the field's archival literature, i.e., its textbooks, journals, conference proceedings, and, nowadays, internet posts and the like, clearly reflect an understanding of that literature as the repository of scientifically authoritative insights about 'people' as they function psychologically under various circumstances—including those circumstances in which they are being viewed as members of collectives such as genders, races, cultures, religious confessions, political affiliations, etc.

Consistent with the sort of thinking encouraged by the present work, however, we begin with a critical appraisal of this long-prevalent view. In point of fact, the vast bulk of the empirical knowledge that has been accumulating in psychology's archival literature for many decades now is constituted of the findings revealed by statistical analyses of quantitatively rendered observations of large numbers of individuals. These 'quantitatively rendered observations' might be assessments of attributes conceived as continuous dimensions, such as intelligence, or might be numerical codes representing categorical assignations such as sex, race,

DOI: 10.4324/9781003375166-2

etc. The large numbers of individuals studied might be regarded either as having been sampled from extant and recognized populations or as representatives of imaginary populations that could, in principle, be created by experimental treatments.[1]

We must ask: does such statistical evidence *actually* shed light on the psychological functioning of individuals? If that *is* so, then exactly *how* is it so? If it is *not* so, then *why not*? These questions are of fundamental relevance to the concerns that will occupy us later in this book, at least if it is assumed that a viable understanding of the social relationships between persons must be founded on a viable understanding of persons themselves.[2]

I will explain in this chapter how mainstream psychology's adoption of aggregate statistical methods of investigation, by now virtually ubiquitous across the field's many subspecialties, has, *de facto,* transformed the discipline from the systematic study of the psychological functioning of individuals that it originally was into what is now, largely, a species of demographic inquiry that I have called *psycho*-demography (cf. Lamiell, 2019). This has happened because the discipline has in fact become almost entirely invested in the production of knowledge about *between-person/group differences,* phenomena that by their very nature can be defined only for real or hypothetical *populations.*

Without question, knowledge of populations is useful for some purposes, and therefore has value in its own right (see, e.g., Lamiell, 2019, chapter 7). However, and contrary to the long-standing tenets of mainstream thinking among social scientists, including but not limited to those self-identifying as 'psychologists,' there is no sound conceptual bridge from knowledge of populations to knowledge of the individuals within, or presumed to be 'represented' by, those populations, whether real or hypothetical. Emphatically, and for reasons I have explained at length and in-depth elsewhere (see Lamiell, 2019, chapter 5), it is *not* the case that a viable conceptual bridge is provided by 'probabilistic' thinking (Banicki, 2018; Proctor & Xiong, 2018; cf. Lamiell, 2018a, 2018b). Alas, the mistakenness of that view is not widely understood within the mainstream of the discipline still commonly referred to as 'psychology,' and, as a result, that view also continues to be falsely propagated beyond the borders of the discipline.

The hurdle created by this epistemic state of affairs must be cleared away if conceptual space is ever to be secured for the consideration of a critically personalistic understanding of persons and their social relationships. Stated more precisely, the task is to make evident (a) why empirical investigations conducted in accordance with the established aggregate statistical methods of contemporary psycho-demographic inquiry fail as an approach to understanding the nature, causes, and consequences of the psychological doings of individuals, and why, therefore, (b) a perspective on inter-personal relationships predicated on psycho-demographic thinking is likewise inadequate for—and even obstructive to—efforts to achieve an understanding of those relationships that might enhance their quality and, in turn, the socio-cultural ethos of community life more broadly.

Clarity on these points will not only help guide those who will conduct psychological research going forward in the revision of their understanding of the knowledge that is actually produced by psycho-demographic studies, but will also, and perhaps more importantly, enable consumers of extant psycho-demographic knowledge to gain a sound critical understanding of its emptiness as a resource for understanding the psychological functioning of individuals. All of this will facilitate the realization of our objectives in later chapters, when the discussion shifts into the domain of the social/interpersonal.

We will approach our task here from a historical perspective, beginning with a brief sketch of a simple experimental procedure illustrative of the approach taken in psychological studies at the discipline's founding as an experimental science late in the 19th century. We will then consider how the gradual displacement of experimental studies of individuals by the method of statistically comparing *aggregates* of individuals transformed the discipline prosecuted by psychology's original experimentalists into something altogether different. Yet despite the obvious radical restructuring of the discipline's investigative procedures that took place during the course of this historic development, what went largely ignored was that the knowledge actually being produced by investigators calling themselves 'psychologists' in fact no longer conformed logically to the original knowledge objectives of the discipline. The few warnings that were eventually sounded concerning the ascendant conceptual mismatch (Bakan, 1955, 1966; Kerlinger, 1979) effected no widespread changes, and renewals of those warnings that have been voiced periodically over the years, though logically sound, have been stubbornly ignored (cf. Lamiell, 2019). The problems resulting from this historic incorrigibility within the field (cf. Lamiell, 2017) continue to be misunderstood or overlooked.

One can only hope that, eventually, this conceptually deplorable state of affairs within 'psychology' will be overcome, so that a widespread appreciation for the fundamental difference between studying individual/group differences in selected psychological doings, on the one hand, and studying the psychological doings of individuals, on the other hand, can take root, just as Stern advised in 1911. Then, but not until, a systematic effort toward reviving the latter in some form suited to 21st-century concerns could begin. In the meantime, those situated outside the field—social theorists, journalists, educators, medical professionals, government officials, etc.—and looking to it for guidance in their own efforts to understand the psychological functioning of human persons can be brought to a sound critical grasp of how little is to be learned in that domain by consulting research findings of an aggregate statistical nature. As the 19th-century German polymath Moritz Wilhelm Drobisch (1802–1896) pointedly advised in 1867:

> It is only through a great failure of understanding [that] the mathematical fiction of an average man ... [can] be elaborated as if all individuals ... possess a real part of whatever obtains for this average person. (Drobisch, 1867, as cited in Porter, 1986, p. 171)

It is precisely that 'great failure of understanding' that initially fueled and now sustains the transformation of psychology into psycho-demography. It is worthwhile to examine a bit more closely what happened.

On the Original Structure of Experimental Psychology

The just-quoted scholar, Drobisch, was a contemporary of Wilhelm Wundt (1832–1920) at the University of Leipzig, in Germany, where, it has long been widely understood, Wundt launched psychology as an experimental science in 1879. As the respective tenures of those two thinkers at Leipzig overlapped, it is entirely possible that they discussed directly with one another the fundamental nature of population-level statistical knowledge. Whether or not this in fact happened, the single most important thing to understand, for our purposes here, about Wundt's 'new science' is that it was not intellectually infected by that 'great failure of understanding' cited by Drobisch.

On the contrary, Wundt and the other leaders of the original experimental psychology[3] were fully cognizant of the discipline's over-arching scientific objective—knowledge of the general laws presumed to govern the psychological doings of human individuals (or, more narrowly at the outset: normal, healthy, adult human individuals; cf. Bakan, 1966)—and of the proper means to that end. Knowledge of what the Belgian polymath Adolph Quetelet (1796–1874) had dubbed, decades earlier, 'the average man' (*l'homme moyen*), even if such knowledge could be regarded as worthy of systematic pursuit in its own right, was not within the epistemic purview of psychology's experimental domain.[4] There, it was clearly understood that experimental results would have to be *fully* defined for *individual* research subjects.[5] It was further understood that knowledge of the generality of those experimental results would be attainable only by examining the results obtained through the study of additional subjects *one at a time.* Let us consider a simple illustration.

An Illustration of Early Experimental Psychology: Basic Experiments in Mental Chronometry

The Dutch scientist F. C. Donders (1818–1889) is widely credited with developing an experimental paradigm, called 'mental chronometry,' that was used in many of the original psychological research laboratories, including Wundt's at Leipzig. Under the terms of that paradigm, an experimental subject would be comfortably situated before an apparatus known as a tachistoscope (ta-KISS-toe-scope, or 't-scope' for short), so as to be able to see, through the instrument's viewing panel, a blank visual field. The subject's positioning would enable comfortable manual manipulation of the keys on the instrument's attached keyboard.

In the simplest phase of an experiment employing visual stimulation, the subject would be instructed to release as quickly as possible a key that he or she was holding down, as soon as a light was detected in the visual field. An attached timer would automatically record the subject's reaction time, measured in milliseconds. Based on several repetitions of this procedure, an experimenter could compute the average of the several reaction times recorded for the subject, and use that average as a good empirical indicator of that subject's true *simple reaction time* (SRT) to visual stimulation by light.

In the next phase of the experiment, the subject would be positioned as described above, and this time instructed to release the depressed key only if the light flashed into the visual field was red. The average of the reaction times obtained over several repetitions of this procedure would be taken as a good empirical indicator of that subject's true *discrimination reaction time* (DRT), i.e., the time needed to both detect the presence of a light and to determine whether or not that light was red.

Theoretically, since the discrimination required in this phase of the experiment, red light versus not-red light is a cognitive operation more complex than that required in the earlier phase, the simple detection of light versus no light, a subject's DRT should exceed by some milliseconds his/her SRT. Moreover, the subtraction of SRT from DRT should provide a good estimate of the subject's true *discrimination time* (DT), i.e., the time required by the subject for the discrimination itself.

In yet a further extension of the experimental procedure, the subject, positioned at the t-scope as before, could be instructed to release one of two depressed keys on the keyboard if the light that appeared was red, but a different key if the light that appeared was yellow. Here once again, the average reaction time recorded over several repetitions of this procedure would provide the experimenter with a good empirical indicator of the subject's true *choice reaction time* (CRT), i.e., the time required to, first, make the required discrimination, red versus yellow, and then to execute the choice of key to release. Extending the rationale of the foregoing, it should theoretically have been the case that *choice time* (CT) would be equal to CRT minus DRT.

Although the assumptions of simple additivity in cognitive operations underlying basic studies in mental chronometry eventually proved empirically unsustainable, our concern here is not with the actual empirical results obtained in that line of inquiry, but rather with the formal features of the experimental procedure itself. In this connection, two points bear special emphasis.

The first of these is that when statistical analyses (such as the computation of average reaction times across several repetitions of a given experimental condition) were performed on the data generated through mental chronometry experiments, all of the data entering into those analyses would have been recordings of the response times of the same individual experimental subject. As indicated above in the description of each phase of a visual reaction time study, *multiple* stimulus presentations with the *same* subject would have been made in order to gain the best estimate of *that* subject's true reaction time to that stimulus presentation. It would not even have occurred to the original experimental psychologists to try to determine what is 'generally' true of normal, healthy adult human subjects regarding SRTs, DRTs, and CRTs to certain visual promptings by averaging the discrete SRT findings across subjects randomly assigned to one experimental condition, the DRT findings across subjects randomly assigned to a second condition, and the CRT findings across subjects randomly assigned to yet a third condition, and then estimating DT and CT values by performing the requisite arithmetic computations based on those average SRTs, DRTs, and CRTs.

The second point to be stressed here is that the *generality* of an experimental result gained with some one subject would have been investigated by conducting identical experiments on multiple individuals, *each considered separately,* and then looking to see the extent to which the separate and fully defined experimental results for each subject could be regarded as effectively equivalent across the individual subjects. Only in this way would it have been possible to establish that some or another outcome was *generally* true in the originally understood sense of *allen gemein,* or *common to all.* Here again: the original experimentalists would not even have considered *averaging* the DTs or CTs across experimental subjects to arrive at proxy empirical indicators of what is *generally* true—in the originally intended sense of *allen gemein*—about the cognitive functioning of normal, healthy adult human subjects.

The 'great failure of understanding' described by Drobisch in 1867 (see above) had yet to infect the fledgling discipline.

The Beginnings of Psychology's Transformation into Psycho-Demography

The Establishment of 'Differential' Psychology

In 1900, a mere two decades after Wundt established his pioneering laboratory, a book authored by William Stern (1871–1938) appeared under the title (in translation) *On the Psychology of Individual Differences: Toward a "Differential Psychology."* Stern (1900) opened his discussion in chapter 1 of that book with words that not only gave the rationale for the new research program that he was proposing, but also underscored what has been said above regarding the knowledge objectives of the original experimental program:

> One of the few features more or less common to all earlier efforts toward a scientific psychology was that the problem was seen as—and only as— a general one. The investigations were concentrated on the most basic elements out of which all psychological life is built up; on the general laws *[die allgemeinen Gesetze]* according to which mental phenomena occur. In this work, the attempt was made to abstract as much as possible from the unending diversity in which we encounter psychological being and living in different individuals, peoples, social classes, genders, etc. The objective was precisely to distill from this broad diversity that which is common *[das Gemeinsame]* ...
>
> Fortunately, we are seeing more and more, in contrast to this view, the emergence of an awareness that the material that has to this point been neglected, specifically, the differential peculiarities of the psyche, deserve attention. (Stern, 1900, pp. 2–3)

The 'differential psychology' Stern proposed in his 1900 book quickly gained popularity among research psychologists, not least because, unlike Wundt-ian style single-subject experiments, the systematic study of individual/group differences seemed to offer possibilities for applying psychologists' expertise in addressing questions arising outside the experimental laboratories, in schools, medical settings, military contexts, and other areas of practical human endeavor. Stern's friend and countryman Hugo Münsterberg (1863–1916) underscored this point in a 1913 publication:

> The study of individual differences itself is not applied psychology, but it is the presupposition without which applied psychology would have remained a phantom. As long as experimental psychology remained essentially a science of the mental laws, common to all human beings, an adjustment to the practical demands of daily life could hardly come in question. With such general laws we could never

have mastered the concrete situations of society, because we should have had to leave out of view the fact that there are gifted and ungifted, intelligent and stupid, sensitive and obtuse, quick and slow, energetic and weak. (Münsterberg, 1913, pp. 9–10)

In the service of this applied psychology, Stern noted in an article published one year after Münsterberg's just-quoted text appeared that

[Wundt's method of experimentation] had to be modified in such a way that it could be applied to larger numbers of persons ... [T]o this end [psychological tests], survey procedures and *methods of statistical analysis made their way into our discipline.* (Stern, 1914, emphasis added)

So rapid was the rise in popularity of the new 'differential psychology' that, just a decade following the publication of his 1900 book, Stern saw the need for an altogether new text on differential psychology, one that, as he explicitly stated on the new work's title page, would take the place of—and hence should not be regarded as—a second edition of the 1900 book. Stern titled the new work (in translation) *Methodological Foundations of Differential Psychology* (Stern, 1911), and in it, he distinguished more precisely and explicitly than he had in 1900 the study of *individuals* from the study of *individual differences.*

Studies of the latter sort, Stern (1911) pointed out, actually yield knowledge of the *attribute variables* according to which individuals or groups have been differentiated from one another within populations (e.g., with regard to levels of intelligence, achievement motivation, race, sex, age, etc.). In such studies, the individuals themselves are 'only the means for conducting the research, inasmuch as they serve as carriers of the attributes under investigation' (Stern, 1911, p. 318; cf. Lamiell, 2019, pp. 53–57). By contrast, Stern argued, genuine studies of individualities would entail not the comparison of individuals with one another along attribute dimensions or categorical differentiations presumed common to them all, but rather, for each individual as such, the specification of the most significant and consequential features of his or her psychological doings.[6]

Unfortunately, the subtle but crucial distinction drawn by Stern (1911) between knowledge about the psychological doings of particular individuals, on the one hand, and knowledge about individual differences in selected psychological doings, on the other hand, was lost on many of Stern's most prominent disciplinary peers (most notably E. L. Thorndike, 1874–1949, and the aforementioned Hugo Münsterberg), and, in turn, throughout succeeding generations of differential psychologists (cf. Lamiell, 2019). Among the most influential 'second generation' differential psychologists of the 20th century, including Anne Anastasi (1908–2001) and Leona Tyler (1906–1993), the population-level statistical examination of variables marking individual/group differences came to be *equated* with the study of individuals, and the knowledge objectives of differential psychology *equated* with the knowledge objectives of the original experimental psychology (see, e.g., Anastasi, 1937, p. vi). With these developments, the unwitting transformation of psychology into psycho-demography was well underway.

Experimental Psychology's Gradual Transformation into a Species of Differential Psychology

Precisely because Wundt-ian style laboratory studies of the psychological doings of individuals were so fundamentally different from systematic non-laboratory studies of individual differences in psychological doings, Stern emphasized that the 'differential' psychology that his 1900 and 1911 books launched would *complement,* and by no means *replace,* the experimental psychology as it existed at that time. However, the rapid and widespread embrace by research psychologists of the investigative methods of differential psychology meant that scientific psychology was becoming more and more a discipline generating strictly correlational knowledge of relationships between variables marking between-person differences that had arisen outside the laboratory (cf. quotation of Münsterberg, 1913 above), and was losing its status as a laboratory science employing controlled experimentation to reveal cause-effect relationships (Danziger, 1990).

There was, however, another form of experimentation extant at the time and in use in other disciplines—most notably in medicine—that seemed to offer experimental psychologists an option for securing the practical advantages of the large-scale statistical methods of inquiry proper to differential psychology *without* surrendering the epistemic leverage provided by systematic experimental manipulations upon which to base cause-effect inferences. Reference here is to *treatment group* experimentation.

In the simplest (and still prototypical) form of such experimentation, individuals are sampled representatively from available research subject pools, and assigned at random to one of two or more experimental treatments. For example, in a study of the relative effectiveness of two different drugs for lowering high blood pressure, some subjects would be assigned at random to treatment drug A, and others to treatment drug B.[7] The central question posed in the study would be addressed by statistically comparing the average post-treatment blood pressures across the two treatment groups created for the specific purposes of the experiment. A statistically significant difference between the groups would be seen as scientific evidence that that difference had been brought about—i.e., *caused*—by the differing treatments.

Despite the stark contrast between this form of experimentation and the original Wundt-ian single-subject approach described earlier, there was, during the 1920s and 1930s, no sustained critical discussion within psychology of the epistemic implications of the shift from single-subject to treatment group experimentation. On the contrary, the view was widely and uncritically adopted that even in the face of this shift, psychology could maintain its status as a discipline for producing knowledge about the psychological doings of individuals. If only by default, the shift under discussion here was seen as nothing more than the paradigmatic replacement of one means for producing such knowledge by another, and in many respects better, means *for producing the same sort of knowledge.* This view could only be maintained by interpreting the patterns revealed by aggregate, population-level statistical analyses of data as indicative of the dynamics of individual-level happenings (Danziger, 1990)—i.e., by indulging Drobisch's 'great failure of understanding' (refer above).

To illustrate, consider the hypothetical drug-effectiveness experiment just described. Let us suppose that it revealed an average post-treatment blood pressure among the subjects exposed to drug A that was 10 points below the average pre-treatment baseline among all of the subjects in the experiment, while the average post-treatment blood pressure among the subjects exposed to drug B fell 5 points below the overall average pre-treatment baseline. If found to be statistically significant (see below), this difference of 5 points in blood pressure reduction would be understood to have been brought about—i.e., *caused*—by the differing drug treatments.

But what is also true, and of deeper relevance to the specific point being made here, is that the 10-point 'reduction effect' evidenced by the *average* result among 'drug A subjects' would be regarded as having been realized in *each one* of those subjects. Similarly, the 5-point 'reduction effect' evidenced by the *average* result among 'drug B subjects' would be regarded as having been realized in *each one* of those subjects. Although few—perhaps not even any—experimental psychologists would recognize themselves as adhering to this 'each one' interpretive proviso, its implicit adoption is built into the presumption of epistemic continuity between treatment group experimentation and the Wundt-ian style single-subject experimentation that preceded it. That is, what an investigation has established as *true on average* must be interpreted as *true in general* if treatment group experimentation is to be regarded as an alternative to Wundt-ian single-subject experimentation as a means to knowledge about individuals.

Be all of this as it may, treatment group experimentalists would not feel compelled to regard the 'each one' proviso contra-indicated by a finding that (to continue the above example) the final blood pressure readings of subjects treated by the same drug, A or B, were not all exactly 10 (or 5) points below baseline. On the contrary, any empirical inequalities in final blood pressure readings among subjects treated by the same drug would be seen as evidence of factors affecting individuals' final blood pressure readings in addition to or in opposition to *but not instead of* the effects, presumptively common to them all, of the particular treatment drug to which each had been exposed.[8] Those further unexamined effects would be understood to have canceled each other out in the data analysis thanks to the random assignment of the subjects to drug treatment conditions. This, presumably, is what would allow the true causal effects of the respective treatment drugs to 'shine through' in the experiment's final results, and be seen as *general* in the sense of *allen gemein,* or *common to all* of the subjects in the respective treatment groups.

If the difference between the respective average final blood-pressure readings for the two treatment groups proved sufficiently large relative to the difference that could have been expected to occur by chance alone (estimated based on the extent of variation in blood pressure readings among subjects who had been exposed to the same treatment), the results of the experiment would be regarded as documenting a 'statistically significant' difference between the two treatment group averages, and the experiment would be so written up for publication in the archival literature.

The foregoing, in a nutshell, describes how the simplest and still most frequently used form of treatment group experimentation works. Note that, vastly unlike Wundt-ian style, single-subject experiments, the eventual widespread reliance on treatment group experimentation has effectively made such experimentation into a branch of differential psychology. In its earlier and strictly correlational incarnation, the focus of differential psychology was on statistical co-variations among variables marking between-person/group differences that had arisen *outside* the research laboratory, presumably as a result of some combination of 'nature' and 'nurture.' In its later-developing experimental incarnation, the research focus came to be on statistical co-variations among variables marking between-person/group differences created *inside* the laboratory by means of experimental manipulations. In either case (and just as well in hybrids of the two), the fundamental nature of the research has always been the same: it entails the generation of knowledge about variables marking individual/group differences in psychological doings, and not knowledge about the psychological doings of individuals. The distinction here is crucial.

Understanding Why Knowledge about Individual Differences in Psychological Doings Is Not Knowledge about the Psychological Doings of Individuals

Brief critical reflection lays bare the logical—but admittedly subtle, hence easily overlooked—fact that individual/group *differences*, which, to re-emphasize, have become the focus of inquiry in virtually all psycho-demographic work, both correlational and experimental, have no empirical existence for any one individual.[9] Such differences are definable as objects of empirical investigation only for collections of individuals. A minimum of two is required, and sound statistically based inferences routinely require many more than that. Just as Stern's (1911) differential psychology text clearly indicates, in studies of individual/group differences, individual subjects serve as nothing more than carriers of discrete levels or categories of the attribute variables that are the actual foci of interest. The knowledge generated through such studies is population-level knowledge of the attribute variables as such, and not knowledge of any individual carrier of some level or category of those variables.

Alas, for the better part of 100 years now, mainstream thinking within 'psychology' has indulged interpretations of aggregate, population-level statistical knowledge as if it were somehow translatable as individual-level knowledge about the members of the studied populations. I have discussed this history at some length elsewhere (Lamiell, 2019), and will not re-trace it here. One point relevant to that discussion warranting particular (re-)emphasis here, however, arises in connection with the concession routinely voiced by contemporary psycho-demographers to the empirical reality that 'there will always be individual exceptions' to the rules articulated on the basis of population-level statistical regularity. This concession seems to reflect recognition that no such regularity, however well-documented empirically, can be presumed to accurately capture the doings of *every* individual member within or putatively represented by the population for which that regularity has been discovered.

On its face, this concession projects proper and well-advised scientific modesty. In actuality, however, the concession is hollow, acknowledging less than it seems to and far less than it must. In fact, no 'rule' reflecting a population-level statistical regularity can be presumed applicable to *any* individual within or putatively represented by the designated population. However startling one might find this claim to be, its truth can be seen in the simple and utterly concrete fact standing in the way of each and every single attempt to use a statistical regularity established for a population as the basis for inferring something about a specific individual case: *this* case 'now' under consideration might very well be one of the ever-possible 'exceptions to the rule.' There is no logical or empirical basis for claiming to know otherwise short of going back to examine that particular case, and it is the exceptionless collision with this epistemic reality, in individual case after individual case, that compels recognition of the fact that population-level statistical regularities are utterly uninformative about individual cases.[10] In practice, one *always* has to 'check to see' whether or not some particular case conforms to expectations based on the interpretation made of some aggregate-level statistical finding. This obdurate reality is what warrants the claim that population-level knowledge is, quite literally, knowledge of *no one.*

To be sure, the proper investigation of an individual case, once it has been carried out, could reveal conformity of aspects of that case to expectations based on some established population-level statistical regularity. The spelling test performance of an individual fourth-grade pupil could, after all, match the average performance within the entire class of pupils. However, (1) the independent investigation of the individual case would always be necessary in order to establish any such individual-aggregate conformity—an individual pupil's spelling test performance would have to be known in order to determine if that performance did or did not match the class average—and (2) even if such conformity were established in this or that individual case, knowledge of *that* fact would add nothing to the already-secured knowledge about the individual in question. This latter bit of knowledge would remain whatever it had been found to be *whether or not* it was then found to conform to expectations based on some aggregate reality. With specific regard to gaining knowledge about an individual, then, the investigation of individual-aggregate conformity is at best superfluous and at worst diversionary.

When all is said and done, the above considerations suffice to reveal why population-level statistical knowledge about variables marking individual/group differences in psychological doings never is, nor can it ever substitute for, knowledge about the psychological doings of individuals. This reveals, in turn, why the now extant body of knowledge that I have branded 'psycho-demographic' fails as a guide to understanding the psychological doings of persons. Clearly, a paradigmatic revival of single-subject inquiry by aspiring psychologists is going to be necessary in order to secure the future of a genuinely psychological science. In the immediately following chapters, an indication of how William Stern's *critical personalism* could guide that revival will be found. Readers will then be well-positioned to focus attention in Part II on a consideration of how critically personalistic thinking could serve to guide the framing of inter-personal relationships and the establishment of an aligned socio-cultural ethos.

Notes

1 For example: consider a simple experiment designed to determine the relative effectiveness of two methods for teaching spelling to fourth-grade pupils, with half of the pupils designated at random for exposure to method A and the other half for exposure to method B. By the rationale for statistically comparing the final average spelling performances of these two groups, the results would be seen as providing a 'window' onto what the average performance of fourth-graders would be in an imaginable population of fourth-graders taught spelling by one or the other of the different instructional methods examined in the experiment.

2 However widely this assumption might be shared, it has not gone unquestioned. On the contrary, a thoughtful and challenging alternative perspective has been developed by the long-prominent social psychologist Kenneth J. Gergen (b. 1935). He argues that it is the realm of the interpersonal that is fundamental and that focusing on the psychological doings of individual persons should be de-prioritized. For my direct, critically personalistic commentary on Gergen's argument as he has developed it in the consideration of moral action (Gergen, 1992), see Lamiell (1992).

3 I am referring here to such scholars as Gustav Fechner (1801–1887), Hermann Ebbinghaus (1850–1909), and E. B. Titchener (1867–1927). There were, of course, many others.

4 Wundt recognized the need for a non-experimental cultural or anthropological psychology that would exist alongside of and complement the experimental psychology. Within that *Völkerpsychologie* (Wundt, 1912), population-level methods of investigation would have a rightful place.

5 Under the terms of the aggregate/statistical methods of psycho-demographic inquiry, experimental *findings* are, of course, obtained for individual research subjects, else there would be nothing for an investigator to subsequently aggregate. But unlike in single-subject experiments, the *results* of psycho-demographic investigations are not determined by the findings obtained with any individual subject. Instead, the results of the investigation remain unknown until the findings for *all* of the individual subjects have been established, aggregated, and statistically analyzed.

6 Stern saw through and past the eventually dominant and still prevailing—but ever misguided—notion that it is impossible to empirically specify the salient features of an individual's psychological doings without comparing him or her with others (see, e.g., Epstein, 1983; Kleinmuntz, 1967; cf. Lamiell, 1987, 1997, 2003, chapter 9, esp. pp. 246–249). The reader will also find relevant to this point my English translations of several of Stern's writing (cf. Lamiell, 2021).

7 In some studies, a non-treatment control group C might be included as well.

8 The reader may note that under this understanding of experimental findings, the 'each one' proviso in the interpretation of causal effects is immune from empirical challenge. In a word, the understanding is *incorrigible.*

9 The trope is hereby challenged that 'everyone is different.' On the contrary, no *one* can be *different*, even if it is always possible for an observer of *two* individuals to identify some difference between them. This point will be discussed more thoroughly in Chapter 6.

10 In the book cited above, I have included an in-depth explanation for why the epistemic gap here cannot validly be bridged by the invocation of probabilistic thinking.

References

Anastasi, A. (1937). *Differential psychology: Individual and group differences in behavior*. New York: Macmillan.

Bakan, D. (1955). The general and the aggregate: A methodological distinction. *Perceptual and Motor Skills, 5*, 211–212.

Bakan, D. (1966). The test of significance in psychological research. *Psychological Bulletin, 66*, 423–437.

Banicki, K. (2018). Psychology, conceptual confusion, and disquieting situationism: Response to Lamiell.*Theory and Psychology, 28*, 255–260. 10.11 77/0959354318759609.

Danziger, K. (1990). *Constructing the subject: Historical origins of psychological research*. New York: Cambridge University Press.

Epstein, S. (1983). Aggregation and beyond: Some basic issues in the prediction of behavior. *Journal of Personality, 51*, 360–392.

Gergen, K. J. (1992). Social construction and moral action. In D. N. Robinson (Ed.), *Social discourse and moral judgment* (pp. 9–27). New York: Academic Press.

Kerlinger, F. N. (1979).*Behavioral research: A conceptual approach*. New York: Holt, Rinehart, & Winston.

Kleinmuntz, B. (1967).*Personality measurement: An introduction*. Homewood, IL: Dorsey Press.

Lamiell, J. T. (1987). *The psychology of personality: An epistemological inquiry*. New York: Columbia University Press.

Lamiell, J. T. (1992). Persons, selves, and agency. In D. N. Robinson (Ed.), *Social discourse and moral judgment* (pp. 29–41). New York: Academic Press.

Lamiell, J. T. (1997). Individuals and the differences between them. In R. Hogan, J. A. Johnson, & S. Briggs (Eds.), *Handbook of personality psychology* (pp. 117–141). New York: Academic Press.

Lamiell, J. T. (2003). *Beyond individual and group differences: Human individuality, scientific psychology, and William Stern's critical personalism*. Thousand Oaks, CA: Sage Publications.

Lamiell, J. T. (2017). The incorrigible science. In H. Macdonald, D. Goodman, & B. Becker (Eds.), *Dialogues at the edge of American psychological discourse: Critical and theoretical perspectives* (pp. 211–244). London: Palgrave-Macmillan. 10.1057/978-1-137-59096-1_8

Lamiell, J. T. (2018a). From psychology to psycho-demography: How the adoption of population-level statistical methods transformed psychological science. *American Journal of Psychology, 131*, 471–475.

Lamiell, J. T. (2018b). Rejoinder to Proctor and Xiong. *American Journal of Psychology, 131*, 489–492.

Lamiell, J. T. (2019). *Psychology's misuse of statistics and persistent dismissal of its critics*. London, UK: Palgrave-Macmillan.

Lamiell, J. T. (2021). *Uncovering critical personalism: Readings from William Stern's contributions to scientific psychology*. London, UK: Palgrave-Macmillan.

Münsterberg, H. (1913). *Psychology and industrial efficiency*. Boston and New York: Houghton-Mifflin.

Porter, T. M. (1986). *The rise of statistical thinking: 1820–1900.* Princeton, NJ: Princeton University Press.

Proctor, R. W., & Xiong, A. (2018). Adoption of population-level statistical methods did transform psychological science but for the better: Commentary on Lamiell (2018). *American Journal of Psychology, 131,* 483–487.

Stern, W. (1900). *Über Psychologie der individuellen Differenzen (Ideen zu einer "differentiellen Psychologie") [On the psychology of individual differences: Toward a "differential psychology].* Leipzig: Barth.

Stern, W. (1911). *Die Differentielle Psychologie in ihren methodischen Grundlagen [Methodological foundations of differential psychology].* Leipzig Barth.

Stern, W. (1914). Psychologie [Psychology]. in D. Sarason (Ed.), *Das Jahr 1913: Ein Gesamtbild der Kulturentwicklung* (pp. 414–421). Leipzig: Teubner.

Wundt, W. (1912). *Elemente der Völkerpsychologie.* Leipzig: Alfred Kröner Verlag.

2 Critical Personalism

Its Core Philosophical Tenets

It was noted in the preface to this volume that a significant prompting for the project was supplied by my involvement in many group discussions in which someone has averred that the tensions arising from racism or sexism or some other problematic aspect of social relationships could be greatly mitigated (even if not necessarily eliminated altogether) if only individuals devoted more effort toward getting to know (and treat) other individuals *as persons*. Though I have yet to witness any opposition to this sentiment, I have also yet to witness any sustained effort toward answering the question of just what 'getting to know and treat' other individuals as persons would entail. A clear answer to this question requires a viable conception of persons, and it cannot be assumed that this is something we all already have and share. The present chapter addresses this concern.

The discussion will proceed in accordance with that worldview, or comprehensive system of thought, that the German philosopher and psychologist William Stern (1871–1938) developed under the name *critical personalism*. Stern's systematic exposition of that system of thought was set forth mainly in three volumes first published, respectively, in 1906, 1918, and 1924 (Stern, 1906, 1918, 1924). Unfortunately, critical personalism remains largely unknown, both among professional scholars and among thoughtful laypersons. Undoubtedly, one reason for this is that Stern wrote in German, and little of his *oeuvre* has been published in English translation (but see Lamiell, 2021). Another reason is certainly that Stern's views were, philosophically speaking, quite out of step with the lines of thought that gained the most widespread attention in 20th-century psychology, and in some key respects have continued to dominate well into the 21st century (Lamiell, 2010, 2013). In his appreciation of Stern following Stern's death in March of 1938, the American psychologist Gordon Allport (1897–1967) wrote that

DOI: 10.4324/9781003375166-3

it troubled [Stern] relatively little that his formulations ran counter to the trend of the times, particularly in American thought .... [H]e believed so intensely in the liberating powers of personalistic thought that he had faith in its ultimate acceptability to others.

Thinking personalistically, Stern became a monumental defender of an unpopular cause. (Allport, 1938, p. 773)

The present work has been written in the conviction that Stern's belief in 'the liberating powers of personalistic thought' was justified, and in the hope that the time for such thought, which Allport (1938) confidently forecast would one day come, has at long last arrived.[1]

An Important Cautionary Note: Personalism Is Not Individualism

The very label '*personal*-ism' might well predispose a reader with no prior familiarity with this particular *-ism* to expect an encounter with some variety of '*individual*-ism.' Sensitive to this possibility, Stern himself made an explicit effort to foreclose it. In the foreword to the second of the three volumes cited above, the 1918 volume titled *The Human Personality,* Stern wrote:

[C]ritical personalism is as distant from a one-sided individualism, which recognizes only the rights and happiness of the individual, as from a socialism, in which individual uniqueness and freedom are choked by the pressure of supra-personal demands. (Stern, 1918, p. x)

The stance adopted by Stern here is quite consistent with the account given by Burgos (2018) of the context in which personalism in general emerged, namely, as a way of thinking that contrasted with individualism, on the one hand, and collectivism on the other.

Specific examples of the non-individualistic spirit of Stern's critical personalism will be introduced in due course as the present work unfolds. For now, the important point is to warn against a premature misjudgment of personalism as a system of thought within which the interests and concerns of the individual are prioritized over those of the community.

The Concept of the Person: Rudimentary Considerations

William Stern completed his doctoral studies in Berlin, at the institution now known as the Humboldt University, under the guidance of the well-known and highly regarded experimental psychologist Hermann Ebbinghaus (1850–1909). Stern's mentor was quite persuaded that the idea of the absolute and inevitable subjection to law of all mental processes ... forms the foundation of all serious psychological work ... . In order to understand correctly the thoughts and impulses of man, we must treat them just as we treat material bodies, or as we treat the lines and points of mathematics. (Ebbinghaus, 1908, pp. 6–9)

Although Stern admired Ebbinghaus and was most grateful for the training in experimental psychology that Ebbinghaus provided him (cf. Stern, 1930), the conceptual bedrock of Stern's thinking was a distinction that positioned him diametrically opposite the conviction expressed by Ebbinghaus in the above quotation, namely, the distinction between *persons* and *things*. Stern found epistemically untenable the thesis that 'basic' psychologists could achieve true knowledge of persons by conceiving of them as mere things. Further, he found morally problematic the proposition that 'applied' psychologists could justify intervening in the lives of persons by treating them as if they were mere things. Stern expressed these concerns poignantly in the first of the three volumes cited above:

The impersonal [mechanistic] natural scientist sits at his desk and writes: 'What we call a human is physically nothing but an aggregate of atoms or, as the case may be, energy quanta, psychologically an aggregate of consciousness contents; nothing happens with him except that which must occur as a consequence of the blind causal relationship of physical elements; and his so-called psychological life is nothing but the mechanical coupling of those physical elements. But then he steps into the nursery, where his child is lying ill; he braces himself against the thought of losing this beloved being—beloved being? What is it about atom + atom + atom (or energy + energy + energy) and idea + idea + idea that merits love? And: lose? In one case (that of the so-called living individual) the elements are bound more tightly to one another; in the other instance [the deceased body] they stand in looser relationship. In the former instance, the influx and outflow of energy is equal, in the latter case it is not. How can this entirely indifferent variability of purely spatial constellation or energy flow mean the difference between joy and despair? And if his other child comes home from school with poor grades, he warns the child: 'You should do better!' Better? Is there a better or a worse in the indifferent coupling of indifferent atoms and indifferent ideas? Whence this scale of value all of a sudden? Whence values at all? And: you should? Where mechanical laws and nothing else are at work, how can there be such a thing as 'should'? Because 'should' signifies nothing other than a determination of one's own doings through the consciousness of a goal. 'Should' is something that can exist only for a self-activated being. A mere string of elements cannot 'should.' (Stern, 1906, p. 79, parentheses in original, brackets added)

The considerations Stern expressed in this passage made obvious to him the need for a clear and uncompromising distinction between persons and things, and, *contra* his mentor, Ebbinghaus, Stern believed that that distinction could be made without undermining the very project of a scientific psychology (cf. Lamiell, 2003). Below, we examine more closely how Stern formulated the person-thing distinction, and, in turn, how he elaborated his conception of the person.

What Is a Person? On the Nature of Personal Being from a Critically Personalistic Perspective

In the same text from which the above quotation was taken, Stern wrote:

A person is an entity that, though consisting of many parts, forms a unique and inherently valuable unity and, as such, constitutes, over and above its functioning parts, a unitary, self-activated, goal-oriented being ... . A thing is the contradictory opposite of a person. It is an entity that likewise consists of many parts, but these are not fashioned into a real, unique, and inherently valuable whole, and so while a thing functions in accordance with its various parts, it does not constitute a unitary, self-activated and goal-oriented being. (Stern, 1906, p. 16)

Although in this passage Stern used the term 'parts' (*Teile*) in reference both to persons and to things, he later eschewed the use of that term in reference to persons, convinced that the term left room for the understanding that the various features of persons (e.g., their respective personality characteristics) were somehow ontologically prior to persons themselves. Stern's convictions as a critical personalist were quite the opposite: whole persons are ontologically prior to any and all of their so-called parts. To better reflect that conviction, Stern eventually settled on the expression 'moments' (*Momente*) as the best way to refer to the many distinguishable facets of personal being (Lamiell, 2021, chapter 1; cf. Stern, 1930). So while things can be said to have 'parts' or 'components' or 'elements' which do exist separately from the things of which they can at some point become parts, a person is otherwise. A person is more properly said to have 'moments' (or 'facets;' my term) which at various times and under various circumstances can be more or less prominent, but which do not exist apart from the person who manifests them.

From very early on in his scholarly life, Stern was concerned about the danger that the still fledgling experimental psychology, modeled as it was on the natural sciences (chiefly physics and chemistry) would inappropriately and needlessly project a thoroughly mechanistic and hence altogether *im*personal understanding of human doings, and he firmly resolved to oppose such an understanding. He wrote in a July 1900 letter to his friend, the University of Freiburg philosopher Jonas Cohn (1869–1947), that what was needed was 'a comprehensive world-view' based on the person-thing distinction; one that would be 'anti-mechanistic' and would temper the then-ascendant 'natural science dogma.' He went on to acknowledge in his letter to Cohn the enormity of the task he had set for himself, but vowed to 'work on it as I can' (cf. Lück & Löwisch, 1994, p. 33).

With the cornerstone of his worldview in place in the form of the person-thing distinction, Stern proceeded to elaborate on his conception of the person.

A Person Is a Unitary, Psychophysically Neutral Being

The mechanistic impersonalism mentioned above was not the only other competing conception of persons extant during Stern's time. Rather, there was also a conception referred to by Stern as 'naive personalism.' That conception qualified, indeed, as a 'personalism' in that it entailed an understanding of personal doings as being directed by an individual's mental faculty rather than brought about by entirely impersonal causal forces operating on physical faculties in strictly mechanical fashion, and subsumable under the categories of biological 'nature' and environ-mental 'nurture.'

It was the philosophically troublesome mind-body dualism entailed by this understanding and commonly traced to the French philosopher René Descartes (1596–1650) that made it what Stern labeled a 'naive' person-alism,[2] and he understood 'critical' personalism as a framework within which that dualism could be avoided without resorting to a mechanistic impersonalism. A key tenet here was the thesis that the person *qua* person functions not as a compartmentalized entity comprised of an immaterial mind or psyche that somehow directs the physical body, but rather as a unitary and hence *psychophysically neutral* entity. As Stern wrote:

It is not that there are the physical and the psychological, but rather that there are real persons. That is the primary fact of the world, [and] the distinction between the psychological mind and the physical body is a fact of secondary order. (Stern, 1906, pp. 204–205)

Underscoring this same point in a later writing, Stern wrote:

[T]he portrayal of the person in mental as well as physical terms is sensible only as a consequence of the existence and being of the person as a psychophysically neutral entity. (Lamiell, 2021, p. 54)

As the reader may already have discerned from the above quotations, endorsing the critically personalistic thesis of psychophysical neutrality does not require one to deny that an understanding of personal doings can be facilitated at various points by the distinction between the mental and the physical. The contention is that these two are moments or facets of the doings of a unitary, whole, and hence most basically psychophysically neutral entity. In order, for example, to understand in critically personalistic fashion even something apparently so simple as a short vocal utterance, it must be grasped not simply as 'the' mind directing the production of words made audible by the workings of a physical voice mechanism activated by neural impulses emanating from the brain, but instead as the expression of meanings—some consciously intended, others perhaps not—via the exercise of one's capacities for vocalization, capacities that, depending upon circumstances, may be deliberately modulated by the whole, unitary person through adjustments in tone, volume, and perhaps even by choreographed facial expressions and other aspects of what is commonly known as 'body language.'

For the analytic and/or didactic purposes of a philosopher, an empirical scientist, a journalist, a cleric, or a thoughtful layperson, a speech act (or any other human doing) might be regarded in a way that separates its mental moments from its physical moments, but from a critically personalistic standpoint it is fanciful to suppose that the act itself somehow requires a bridging of that mental-physical separateness. Since in the act itself there is no such separateness, there is no bridging involved or in need of explanation. In other words, the act itself is properly understood not as the coordinated execution of distinctly mental and physical 'parts,' but as a fundamentally unitary and psychophysically neutral occurrence with mental and physical *moments*.

A Person Is a Causally Effective Agent

From an impersonal standpoint such as that advocated by Stern's mentor Ebbinghaus (refer to quotation of Ebbinghaus above), persons are properly regarded as things. This means viewing persons as strictly passive entities the doings of which are in principle completely explainable in terms of (a) the biochemical dynamics of physical constitution ('nature'), and (b) the forces brought to bear on those physical constitutions from without (environmental 'nurture'). This is not the view adopted by Stern the critical personalist. On the contrary, he made explicit his conviction that 'the person conceptualized critically is causally effective' (cf. Lamiell, 2021, p. 48), and he hastened to add that he intended more by this statement than a simple acknowledgment that wholly impersonal cause-effect sequences are played out within the person *qua* thing, itself functioning essentially as a mere vessel:

> Certainly, the parts contained with a person carry out activities, but these activities do not, in the aggregate, make up a person's doings; they are rather but the raw material, condition, and limits of those doings. The basis for attributing this causal effectiveness to the person is the goal-directedness of an individual's doings … On this view, causal effectiveness and teleology converge within the person.

Stern was fully aware that in embracing a teleological[3] perspective on persons' doings he was charting a course for psychological studies deeply incompatible with what he called the 'teleophobia' that, already by his time, had developed within mainstream thinking (cf. Lamiell, 2021, p. 10). Nevertheless, he was firmly convinced that 'the idea of purpose is the very key to a true understanding of personal being' (Lamiell, 2021, p. 10).

It is important to note here that Stern did *not* stipulate that a person's conscious awareness of the purpose(s) served by some doing is a prerequisite for the causal effectiveness of the purpose(s) in bringing about that doing. On the contrary, Stern saw clearly that much that is causally effective in bringing about a person's doings can transpire outside the range of that person's conscious awareness at some particular point in time.

The second point bearing mention here is that in construing persons as causally effective agents, the critical personalist endows teleologically functioning persons with a certain latitude of *freedom* in their personal doings. This is another reflection of the conviction that personal doings are not fully determined by extra-personal forces of biological nature and environmental nurture. This presumption of freedom has been problematic for mainstream thinking in psychology at least from the time that the discipline was established as an empirical science. In fact, this is what Stern meant by his reference to 'teleophobia.'

Stern was firmly of the view, however, that persons not only *are* validly regarded as possessing a certain latitude of freedom, but *must* be so regarded in order for there ever to be a firm conceptual basis for judging the rightness or wrongness of some personal doing. As he put it in the lengthy quotation cited early in this chapter, 'a mere string of elements (supplied by nature and nurture) cannot "should"' (parentheses added here). The argument Stern mounted for including the notion of personal causation in the scientific study of personal doings will be discussed in further detail below (see also Lamiell, 1992; 2021, pp. 55–60).

A Person Is an Inherently Evaluative Being

If persons are themselves causally effective agents in the pursuit of certain goals, ends, or purposes, then the question arises as to the basis upon which those goals, ends, or purposes are set. Who or what determines them? Stern's critically personalistic answer to this question is persons themselves.

In Volume III of *Person and Thing,* published in 1924 under the title 'Philosophy of Value,' Stern wrote *Ich werte, … also bin ich Wert* (Stern, 1924, p. 35). Arguably, the most direct and literal translation of this sentence would be 'I evaluate, … therefore I am value.' However, the sentence is pregnant with further profound implications for the concept of 'person.'

The process of evaluation, i.e., e-valuation, involves projecting values outward, onto entities or circumstances or developments in one's world. As Stern phrased matters in many other places in his writings, persons '*strahlen Werte aus,*' meaning that they radiate or project values outward into the physical and inter-personal world around them, in a fashion analogous to the way in which the sun radiates heat and light. If this is so, then values must in some fundamental sense inhere within persons to begin with. Since this is not true of things, which can only be the passive targets of evaluations, and since all entities in the world are either persons or things, it must be the case (at least from a secular standpoint) that the evaluation process originates in persons.[4] That is: values inhere in persons, making persons *inherently value-able.* It is but a small conceptual step from here to the thesis that persons—in contrast to things—are inherently *valuable.* A mechanistic, impersonal worldview denies this thesis, and from a critically personalistic standpoint is problematic in this regard both epistemically and morally: epistemically in that it *falsely portrays* persons as things; morally in that it *sanctions the treatment* of persons as things.

In ways that will be elaborated in subsequent chapters, these considerations have important implications for our understanding of how social engagements and cultural practices would be re-shaped on the basis of critically personalistic thinking.

A Person Is a Distinctive Individuality

As noted in Stern's definition of 'person' cited toward the beginning of this chapter, critical personalism posits the inherent value of persons, as just explained, but also each person's *distinctiveness* as an individual. and their *wholeness,* i.e., their unitary nature. He wrote in *The Human Personality* that

> Despite all commonalities through which persons are like other instances of humanity, representatives of a race, members of a gender, despite all broader and narrower lawful regularities that are at play in all personal happenings, there ever remains an ultimate reality, according to which every person stands as a world unto oneself before every other person. (Stern, 1923, p. 7; see also Lamiell, 2021, pp. 49–50)

From the perspective of critical personalism, the distinctiveness of an individuality is revealed, both to the person him/herself and to others, by what is embraced versus rejected by the person in the realm of human values. In ways that may not be entirely obvious to the reader, this perspective on individual distinctiveness differs importantly from the one that has long-dominated thinking in mainstream psychology, according to which the distinctiveness of an individuality hinges not on how it differs from *what it is not but would otherwise be*, but, instead, entirely on how it differs from *other extant individualities*. From that perspective, an individual's distinctiveness is known by contrasting what is the case for that individual with what is the case for other individuals. As Epstein once expressed this view rather forcefully, 'it is *meaningless* to interpret the behavior of an individual without a frame of reference of others' behavior' (Epstein, 1983, p. 381, emphasis added).

Brief reflection suffices to reveal why this claim by Epstein (1983) cannot be true: To say that a person cannot be meaningfully characterized as, say, 'honest' or 'tolerant' unless and until that person is compared with others is to say that, prior to such a comparison, the target individual has no standing at all with respect to the attributes of honesty or tolerance. But if that were true, then no comparison of the target individual with others would ever be possible, for in order for that individual to be judged 'more' or 'less' (or, for that matter, 'equal to') some other(s) with respect to some attribute(s), the target individual must have *some* standing with respect to the attribute(s) to begin with.

Characterizations of individuals based on extant differences between individuals are possible only because meaningful characterizations of individuals prior to and independent of such empirically-based comparisons are rationally possible.[5] In Chapter 6, we will consider the broader implications of this point for a critically personalistic grounding of interpersonal discourse and socio-cultural life. At this point, it suffices to reiterate for emphasis that, from a critically personalistic perspective, (a) the distinctiveness of an individuality is grasped through considerations of *what is versus what is not* the case about that person, and (b) such considerations *might*—but need not—distinguish a given individuality from other extant individualities.

Who Is This Person? On the Basic Dynamics of Psychosocial Development from a Critically Personalistic Perspective

Above, attention was directed to the ideas that, from a critically personalistic perspective, personal *being* is both (a) inherently evaluative, and (b) teleological in nature. This means that personal *doings* must be understood as strivings that, consciously or otherwise, are *directed* toward the achievement of certain ends or goals. Stern elaborated on these basic ideas by distinguishing between various categories of goals *(Zwecke)*. In doing so, he effectively formulated a framework for understanding the psychosocial development of persons.

The most basic category of goals circumscribes those that are the person's *own* goals *(Selbstzwecke)*. In a way that serves to draw attention to the teleological nature of personal doings, Stern labeled this category of goals 'own-goals' *autotelic* ('auto-telic'), and he identified the two major subcategories of such goals as self-'maintenance,' or survival (*Selbsterhaltung*), and self-'unfolding' or development (*Selbstentwicklung*). In positing these autotelic goals as fundamental to personal doings, Stern was projecting a conception of persons as entities naturally valuing not only their own survival, but, beyond that, their own growth, in a way finally akin to the ancient Aristotelian notion of *flourishing* (cf. Fowers, 2015).

Here, however, another indication of the non-individualistic nature of critically personalistic thinking (refer above) surfaces. Stern wrote:

> [T]he person who pursues only his/her own narrow individual goals would be an extensionless-less point in emptiness. Only the goals extending beyond the self give the person concrete content and living coherence with the world. (Stern, 2010, p. 130)

The category of goals Stern named *heterotelic* (hetero-telic or others' goals) include goals that are originally situated in entities external to the focal person (the person under immediate consideration). These external goals *(Fremdzwecke)* include the self-goals of other persons, and when the focal person adjusts his/her objectives to the objectives of one or more other persons, the goals resulting from the adjustment are called *syntelic*.

In other cases, the entities in which the heterotelic goals are situated are supra-individual collectives such as families, ethnic groups, religious congregations, civic organizations, or even a deity. Goals formed as adjustments to supra-individual collectives are called *hypertelic*. Still another possibility is realized when a person adjusts his/her self-goals to broad human ideals such as truth and justice. These adjustments are said to be *ideotelic*.

All of these various kinds of adjustments, Stern wrote, reflect the human capacity

> ... to take up the heterotelic into the autotelic. The former ... are appropriated within and formed according to one's own self. Only in this way does it become possible that the surrender to [external] goals nevertheless does not signify any de-personalization or degradation of the personality into a mere thing and tool, but that, on the contrary, the personality becomes, through its embodiment of those outer goals, in itself-activity, a *microcosm*. (Stern, 2010, p. 131, emphasis added)

Noteworthy here is the fact that, from a critically personalistic standpoint, the distinctiveness of a person (refer to discussion above) is not somehow compromised by that person's adaptation to the goals of others. On the contrary, such adaptation is inherent within the process of *becoming one with* others—the *microcosm* of which Stern wrote.

The process of 'appropriation' identified in the above passage is one for which Stern adopted the technical expression *Introzeption*, or 'int(e) roception.' With this term, he was seeking to capture the idea of critically and deliberately 'taking in,'—as opposed to being passively 'stamped by'—certain other- ('heterotelic') goals, and purposely adjusting one's own goals to them. This process epitomizes what Stern called *person-world convergence:* the melding, as it were, of a person's own goals and objectives with the goals and objectives of entities existing apart from oneself. He wrote:

> From the standpoint of the person, the world is not only at hand as a part of one's goal system (specifically with respect to 'heterotelic' goal setting) but is also *a co-determinant of one's doing and being* ... . I refer to this co-determination as *'convergence.'* ... The person is, according to his/her inner dispositions simultaneously goal-striving *and in need of supplementation.* It is in view of this latter need that a role is postulated for the participation of the world in the person's development. (Stern, 2010, pp. 132–133, emphases added, parentheses in original)

We see vividly reflected in this passage the critically personalistic view of the individual as being naturally in need of and complemented by the outside world in the course of becoming the person s/he is at any given point in time. Stern understood the personal 'dispositions' to which he referred in the above passage as *potentialities,* the realization of which might be facilitated or impeded by the world around oneself, i.e., the person's *Umwelt.* From a critically personalistic standpoint, it is vital to understand that an individual's personal being and doings are not strictly determined, in some narrow and mechanistically causal way, by those factors. The person him/herself is always actively *converging with* the realities of the outside world, and is never fully, decisively, and passively *shaped by* those realities. The goals that the person chooses to pursue, which of course reflect both the values that the person embraces and those s/he rejects, circumscribe the parameters of the person's being and doings at any given point in his/her psychosocial development.

Of course, the constellation of values embraced and rejected by a person can change over time, and, in many instances, this happens as a consequence of realities arising in the outside world. However, this dynamic remains understood within critical personalism as a dynamic of person-world convergence, and not as a dynamic of causal forces emanating from the outside world and blindly determining the course of a person's psychosocial development, akin to the behavior of billiard balls on a pool table. This is the decisive point, and, as we will see later, it has profound implications for the socio-cultural ethos that would arise from the widespread adoption within a community of the critically personalistic worldview.

Notes

1 In addition to his familiarity with Stern's writings, Allport knew Stern personally, and even boarded in the Stern home for some time during the mid-1920s while doing post-doctoral work with Stern at the University of Hamburg. Allport also hosted Stern at Harvard University while Stern taught a summer course there after fleeing Nazi Germany for the U.S. in 1934.

2 Years later, in a 1949 monograph titled *The Concept of Mind,* the British philosopher Gilbert Ryle (1900–1976) would deprecate this presumed special mental component of persons as 'the ghost in the machine.'

3 The word 'telos' is Greek for 'end' or 'goal' or 'purpose,' and this is why accounts for human doings that refer to their ends or goals or purposes are called 'teleological.' See Rychlak (1981) for a much more thorough discussion of this topic.

4 Catholic personalists, of which there have been, historically, a great many (Burgos, 2018; Burrow, 1999), have contended that God is the ultimate source of values and, hence, the ultimate Person. Stern was not Catholic, and did not align critical personalism with any specific religious denomination. His heritage

was Jewish, although he did not engage regularly in Jewish religious practices, and his conception of God was reflected in his pantheistic outlook on the world.

5 For a survey of empirical studies bearing directly on this point, see Lamiell (1987, chapter 6), Lamiell and Durbeck (1987), Lamiell, Foss, Larsen, and Hempel (1983), Lamiell, Foss Trierrweiler, and Leffel (1983). See also Lamiell and Trierweiler (1986).

References

Allport, G. W. (1938). William Stern: 1871–1938. *The American Journal of Psychology, 51*, 770–773.

Burgos, J. M. (2018). *An introduction to personalism.* Washington, D.C.: The Catholic University of America Press.

Burrow, R., Jr. (1999). *Personalism: A critical introduction.* Nashville, TN: Chalice Press.

Ebbinghaus, H. (1908). *Psychology: An elementary text-book* (M. Meyer, Trans.). D. C. Heath & Co. Publishers.

Epstein, S. (1983). Aggregation and beyond: Some basic issues in the prediction of behavior. *Journal of Personality, 51*, 360–392.

Fowers, B. J. (2015). *The evolution of ethics: Human sociality and the Emergence of Ethical Mindedness.* London, UK: Palgrave-Macmillan.

Lamiell, J. T. (1987). *The psychology of personality: An epistemological inquiry.* New York: Columbia University Press.

Lamiell, J. T. (1992). Persons, selves, and agency. In D. N. Robinson (Ed.), *Social discourse and moral judgment* (pp. 29–41). New York: Academic Press.

Lamiell, J. T. (2003). *Beyond individual and group differences: Human individuality, scientific psychology, and William Stern's critical personalism.* Thousand Oaks, CA: Sage Publications.

Lamiell, J. T. (2010). Why was there no place for personalistic thinking in 20th century psychology? *New Ideas in Psychology, 28*, 135–142.

Lamiell, J. T. (2013). Critical personalism: On its tenets, its historical obscurity, and its future prospects. In J. Martin & M. Bickhard (Eds.), *Contemporary perspectives in the psychology of personhood: Philosophical, historical, psychological, and narrative* (pp. 101–123). Cambridge, UK: Cambridge University Press.

Lamiell, J. T. (2021). *Uncovering critical personalism: Readings from William Stern's contributions to scientific psychology.* London, UK: Palgrave-Macmillan.

Lamiell, J. T., & Durbeck, P. (1987). Whence cognitive prototypes in impression formation? Some empirical evidence for dialectical reasoning as a generative process. *Journal of Mind and Behavior, 8*, 223–244.

Lamiell, J. T., Foss, M. A., Larsen, R. J., & Hempel, A. (1983). Studies in intuitive personology from an idiothetic point of view: Implications for personality theory. *Journal of Personality, 51*, 438–467.

Lamiell, J. T., Foss, M. A., Trierweiler, S. J., & Leffel, G. M. (1983). Toward a further understanding of the intuitive personologist: Some preliminary evidence for the dialectical quality of subjective personality impressions. *Journal of Personality, 53*, 213–235.

Lamiell, J. T., & Trierweiler, S. J. (1986). Interactive measurement, idiothetic inquiry, and the challenge to conventional 'nomotheticism.' *Journal of Personality, 54*, 460–469.

Lück, H. E., & Löwisch, D.-J., (Eds.) (1994). *Der Briefwechsel zwischen William Stern und Jonas Cohn: Dokumente einer Freundschaft zwischen zwei Wissenschaftlern* (Correspondence between William Stern and Jonas Cohn: Documents of a friendship of two scientists). Frankfurt am Main: Verlag Peter Lang.

Rychlak, J. F. (1981). *A philosophy of science for personality theory*, second edition. Malabar, FL: Krieger.

Stern, W. (1906). *Person und Sache: System der philosophischen Weltanschauung, erster Band: Ableitung und Grundlehre* [Person and thing: A systematic philosophical worldview, Volume 1: Rationale and basic tenets]. Leipzig: Barth.

Stern, W. (1918). *Person und Sache: System der philosophischen Weltanschauung, zweiter Band: Die menschliche Persönlichkeit* [Person and thing: A systematic philosophical worldview, Volume 2: The human personality]. Leipzig: Barth. (first edition)

Stern, W. (1923). *Person und Sache: System der philosophischen Weltanschauung, zweiter Band: Die menschliche Persönlichkeit,* dritte unveränderte Auflage [Person and thing: A systematic philosophical worldview, Volume 2: The human personality, third unaltered edition]. Leipzig: Barth.

Stern, W. (1924). *Person und Sache: System des kritischen Personalismus, dritter Band: Wertphilosophie* [Person and thing: Systerm of critical personalism, Volume 3: Philosophy of value.] Leipzig: Barth.

Stern, W. (1930). William Stern intellectual self portrait (S. Langer, Trans.) In C. Murchison (Ed.), *A history of psychology in autobiography* (Vol. 1, pp. 335–388). Worcester, MA: Clark University Press.

Stern, W. (2010). Psychology and personalism (J. T. Lamiell, Trans.). *New Ideas in Psychology, 28*, 110–134. 10.1016/j.newideapsych.2009.02.005.

3 The Challenge of Reviving Psychological Studies

Some Further Historical Perspectives and Some Possibilities for Moving Forward

In March of 1904, when William Stern was hard at work on the first of the planned three volumes of *Person and Thing*, a volume he would title *Derivation and Basic Tenets (Ableitung und Grundlehre)*, he wrote in a letter to Jonas Cohn that

> if things go well, I hope to be able to publish the first volume by the end of the year ... . Only very few people will understand it, and hardly anyone will agree with it. It is difficult and quite off the beaten path, but nevertheless I believe in its future. (Stern letter to Cohn, March 9, 1904; cited in Lück & Löwisch, 1994, p. 58)

It happened that volume I of *Person and Thing* would not appear in print until 1906. As Stern had anticipated, however, it would find little resonance within the mainstream of scientific psychology. This, alas, proved to be a harbinger of things to come. Matters would not improve appreciably over the ensuing three decades. During that time, Stern published many additional works, including volumes II and III of *Person and Thing*. Through those various works, he explicated the fundamental commitments of critically personalistic thinking, and showed the relevance of that thinking not only to basic theoretical issues in psychology but also to the efforts of applied psychologists in industry, in medicine, in psychological testing, and in education.[1] Nevertheless, following Stern's passing in 1938, his American colleague Gordon Allport could still write only of his hope that, eventually, Stern's thinking would 'have its day' (cf. Allport, 1938, p. 773; refer to discussion in Chapter 2 of Allport's posthumous appreciation of Stern).

Given my call in the present volume for a revival of psychological studies within a critically personalistic framework, it would seem appropriate to pay due attention here to the enduring challenges that efforts in this direction will face today. That is one major objective of the present chapter. I undertake this effort in the conviction that doing so

DOI: 10.4324/9781003375166-4

will serve the interests of students seeking to become psychological researchers themselves. But I also believe that the discussion in this chapter should prove informative for those who, though not aiming to conduct their own psychological studies, will be consulting the discipline's research findings for needed guidance in their own professional endeavors outside of psychology *per se.* I have in mind here philosophers, sociologists, political scientists, economists, educators, clergy, and journalists. Perhaps most importantly when all is said and done, I have in mind here thoughtful lay persons seeking critical insights into their own and their fellow individuals' doings as human persons.

One—perhaps *the*—major challenge here is that of overcoming the inertness of long-standing and deeply entrenched beliefs about how psychological research is properly carried out 'scientifically.' As explained in Chapter 1, these beliefs are rooted in firm but erroneous understandings of what can be revealed about individuals by the kind of knowledge secured through competently executed statistical studies of populations. The actual limitations of such studies *vis-a-vis* their stated objectives have been pointed out by numerous authors in various works extending over many years (cf. Lamiell, 2019), and yet the false interpretations and unjustified knowledge claims persist.

The difficulties here are of just the sort captured by the quip, widely attributed to Mark Twain, that 'what gets us into trouble ain't what we don't know; it's what we know for sure that just ain't so.' For as long as mainstream thinkers perseverate in their insistence that scientific knowledge about individuals is secured through studies of statistical relationships between variables marking individual and group differences (and therefore definable only for populations of individuals), a perseveration that I have branded *statisticism* (Lamiell, 2019), but for which I am now proposing the shorter and more easily pronounced term *statism* (emphasis on first syllable, rhyming with 'cat': STAT-izm),[2] the prevailing blindness to the very *need* for a revival of genuinely psychological inquiry will persist, and psycho-demographic knowledge about populations will continue to be masqueraded as psychological knowledge about individuals.

In Stern's own day, 'statism,' though nascent, was not nearly as pervasive as it is today. Nevertheless, critical personalism still failed to generate much interest within the mainstream of psychology. The fact of this matter suggests that the obstacles to a contemporary revival of psychological science may lie beyond 'statism' itself.

The essence of the argument to be developed in this chapter is that the one-time discipline of psychology long ago lost touch with its intellectual roots. I will argue that the single biggest challenge facing those who would now work to revive psychological studies will be to critically re-connect with those lost roots, and to then adapt the discipline to 21st-century circumstances and substantive concerns. The challenge is a formidable one, but I believe it can be met. An important first step entails gaining additional historical perspective (beyond that offered in Chapter 1) on the course that has led psychology to its current situation.

From Psychological Science to a Scientistic 'Psychology'

When experimental psychology was launched late in the 19th century, its place in the academy was as a sub-specialty within philosophy. Reflective of this fact is a diary entry that the 19-year-old university student Stern would make some time in 1890:

> The die is cast ... . It is in philosophy that I will either succeed or fail. It comforts me to know that a sub-discipline within philosophy is wide open to me, namely, psychology. (quoted in Stern, 1927, p. 6)

For the duration of his life, Stern would regard the conceptual ties between philosophy and psychology as integral and hence indissoluble. However, already by the turn of the 20th century, when experimental psychology was scarcely even 20 years old, efforts were underway to sever the discipline from philosophy. Many specialists in experimental psychology wanted to focus on the conduct of their empirical investigations in laboratories, and on training their students to do the same. For their part, those scholars who at the time were referred to as 'pure philosophers' *(reine Philosophen)* wanted to concentrate their scholarly and pedagogical efforts in the traditional areas of metaphysics, epistemology, ethics, and the history of philosophy, and not have to bother with the training—first of themselves and then of their students—in the methods of controlled laboratory experimentation and data analysis. Hence, considerable support for a divorce of the two disciplines was present within each of them.

That movement was more than a little troubling to the acknowledged founder of experimental psychology, Wilhelm Wundt, and in a 1913 publication titled (in translation) *Psychology's Struggle for Existence (Die Psychologie im Kampf ums Dasein),* Wundt gave voice to his concerns. The most fundamental of those concerns stemmed from the conviction that philosophical considerations were inextricable from psychological theory. Hence, Wundt argued:

> In a psychology divorced from philosophy, philosophical considerations will be latent, and so it is possible that psychologists who will have abandoned philosophy, and whose education in philosophy will therefore be deficient, will be projecting those considerations anyway,

but [doing so] through an immature metaphysical perspective. As a result of such a separation, therefore, no one will suffer more than psychologists—and, through them, psychology. (Wundt, 1913/2013, p. 204; brackets added)

Wundt went on from there to underscore the warning he had just sounded:

If this matter takes the course that both parties want, philosophy will lose more than it will gain, but psychology will be damaged the most. Hence, the argument over the question of whether psychology is or is not a philosophical science is, for psychology, a struggle for its very existence. (Wundt, 1913/2013, p. 205)

In Stern's last major publication, which appeared more than two decades after the publication of the original German-language work by Wundt just cited, Stern emphasized his own concerns on the matter with an argument strikingly similar to that that had been made by Wundt. Stern insisted that the complete separation of philosophical and empirical concerns within scientific psychology is not possible. 'On the contrary,' he argued, ... a symbiotic relationship between philosophical considerations and [empirical] findings

is unavoidably necessary. The conviction, still now widespread, that psychology could or should become a discipline fully independent of philosophy leads either to a psychology without a psyche or to scientific work that incorporates a world view and grounding epistemological presuppositions that are not consciously recognized. (Stern, 1935, p. 10)[3]

The 'psychology without a psyche' about which Stern warned is precisely the psychology that the founder of stimulus-response psychology, John B. Watson (1878–1958) trumpeted with his 1928 claim, quoted in Chapter 1, that

with the behavioristic point of view now becoming dominant, it is hard to find a place for what has been called philosophy ... . Philosophy has all but passed. (Watson, 1928, p. 14)

The notion that the empirical discovery of scientific regularities would—or even could in principle—render non-empirical, philosophical considerations irrelevant is at the core of an orientation that is known as *scientism* (see Gantt & Williams, 2018), and it is precisely this orientation that Watson was embracing in the passage just quoted. It is, moreover, scientistic thinking as it would apply specifically to empirical psychology from which Stern explicitly distanced himself in the 1906 book cited at the outset of this chapter:

> One ought not to think that the system of thought I call 'critical personalism' has been shaped by my engagements in the domain of scientific psychology. On the contrary: I oppose *psychologism,* [a view that] would subordinate metaphysical considerations, which are simultaneously meta-psychological, to the [empirical regularities revealed by the] science of consciousness. (Stern, 1906, p. viii, brackets and emphasis added)

On Stern's view, the notion that the discovery, through laboratory investigations, of regularities such as those linking the behavioral 'responses' of organisms to eliciting 'stimuli' would render non-empirical, philosophical considerations irrelevant to a coherent scientific *understanding* of those relationships was untenable.

However pointedly Wundt and Stern and a scant few others among their contemporaries warned of the untoward consequences of a psychology-philosophy divorce, there was at the time no stanching the flow in that direction. Indeed, already by 1937, some three years after Stern had been welcomed into a position on the psychology faculty at Duke University in Durham, North Carolina, there were signs of the divorce's *fait accompli*—to be seen, paradoxically, in Stern's statement in a letter to Cohn dated July 16 of that year that he, Stern, was about to become an exception to that trend:

> It will interest you to know that next year I'll also belong to the Department of Philosophy—something that is rare in the U.S. due to the sharp separation of philosophy and psychology. (Stern letter to Cohn, July 16, 1937; cited in Lück & Löwisch, 1994, p. 180)

Stern passed away less than a year after that letter was written, and now, more than eight decades later, we find that the discipline still referred to as 'psychology' is as distant as ever from its 'mother discipline,' philosophy, *and* from its very own roots as scientific psychology. Thus, the consequences of this development have in fact been highly injurious to scientific psychology, just as Wundt predicted they would be.

When all is said and done, the primary function of philosophical analysis is to secure *conceptual clarity* (Bennett & Hacker, 2003). Unfortunately, appreciation among mainstream psychologists for the importance of that aspect of scientific work in concert with empirical investigation has declined over the years, paralleling psychology's increasing distance from philosophy (Machado and Silva (2007). The inevitable consequence of this devaluation of conceptual work is *diminished* conceptual clarity in psychologists' putatively scientific pronouncements—the very consequence foreseen by Wundt and by Stern decades ago. Compounding difficulties here is mainstream psychology's incorrigibility in the face of arguments revealing emergent conceptual confusions (see, e.g., Lamiell, 2017), arguments that are themselves quite obviously—and of necessity—conceptual in nature. Exhibit A in this regard is the perseveration I referred to above as 'statism.' It is arguably the most consequential form of scientism/psychologism present within the discipline today, and it stands as a huge impediment to the successful revival of psychological science.

A case in point: The deep and abiding confusion in 'psychology' between the aggregate and the general

In a short journal article published in 1955, psychologist David Bakan (1921–2004) advised his contemporaries in the field as follows:

> The failure to distinguish between general-type and aggregate-type propositions is at the root of a considerable amount of confusion which currently prevails in psychology. There are important differences in the research methods appropriate to these two types of propositions. The use of methods which are appropriate to the one type in the [attempted] establishment and confirmation of the other leads to error. (Bakan, 1955, p. 211, brackets added)

What had transpired in the discipline of psychology of which Bakan was writing in 1955 was the gradual abandonment of the method of investigating individuals, one at a time (here 'Method 1') in favor of the method of investigating large numbers of individuals simultaneously (here 'Method 2'). This latter method entailed the use of aggregate statistical concepts and procedures to probe the nature, sources, and consequences of differences between individuals and groups, and could be employed whether the between-person/group differences under investigation had arisen prior to the investigation and were assessed by means of tests/category coding or had been created deliberately in the course of investigation by the imposition of different experimental treatments (cf. Lamiell, 2019). Belief in the epistemic continuity of these two investigative approaches—i.e., the belief that Method 2 was an *alternative* to Method 1 as a means of securing knowledge of individual doings, as opposed to something wholly *different* from Method 1 and suited to producing knowledge of a fundamentally different sort—was at the root of the failure of distinction that concerned Bakan (1955). That failure led to the discursive practice of referring to knowledge of what obtains for collectives of individuals *on average* (or, equivalently, *in the aggregate*) as if it were knowledge of what is true *in general* of the individuals within those collectives/aggregates, in the *true-for- each-of-many-individuals* sense of 'general' understood by practitioners of Method 1 (refer to Chapter 1).

Bakan's 1955 article issued from his insight that the assumption of epistemic continuity just described, and the emergent discursive practice regarding what is 'generally' true, was based on confusions that needed to be corrected. Alas, his 1955 advisory was little heeded by his contemporaries. The confusions to which he had quite properly pointed thus continued unabated, and, even more disturbing, they continued even after Bakan reiterated his concerns in a 1966 journal article (Bakan, 1966). There, he emphasized the fact that the objective of empirically validating general-type propositions does not obviate the need to attend to individual-level doings. This is so because, by its very nature, a proposition that is *generally* true of individuals within some circum-scribed set is so precisely and only insofar as it is true for *each one* of the individuals within that set, and there would be no way to establish this except by investigating matters at the level of the individual.

This is not true of aggregate-type propositions. On the contrary: while something that is true *on average* or 'in the aggregate' for a collection of individuals *could* be true of each individual within that collective, it would not necessarily be true of each one of them and *might* not be true of *any* of them.[4]

A revival of psychological studies is going to require the re-establishment of a clear distinction between aggregate-type and general-type propositions, in full accord with the advisory David Bakan sounded nearly 70 years ago. The fact that that advisory has not been heeded up to now is a major reason that psychological studies have been replaced by psycho-demography.

A corrective might have been achieved through, or at least sparked by, a work published in 1979 by Fred N. Kerlinger (1910–1991), a research methodologist highly regarded among mainstream psychologists during his years. The work was titled *Behavioral Research: A Conceptual Approach* (Kerlinger, 1979), and, in line with a point developed earlier in this chapter, the use of the term 'conceptual' in the title might have lessened the work's readership right from the start, relative to other more technical, computationally oriented research methods texts that Kerlinger also authored (cf. Kerlinger, 1986; Kerlinger & Pedhazur, 1974).

In any case, the 1979 book contained a discussion of what Kerlinger termed a 'troublesome paradox' (p. 275) running throughout psychol-ogy's research literature, defined by the fact that while psychological theory is aimed at explicating psychological doings at the level of the individual, the statistical methods that Kerlinger (and most others in the field) believed properly scientific psychological investigation required demanded that hypotheses be formulated and tested at the aggregate level.[5] The question begged by Kerlinger's 'troublesome paradox' was—and remains: how do statistical tests of hypotheses carried out at the aggregate level actually evaluate theoretical propositions about the nature, sources, or consequences of individual doings? The correct answer is: *they don't*—just as Bakan (1955, 1966) had warned the field many years earlier (refer above). Alas, there is no indication in Kerlinger's, 1979 text that he had given any attention to Bakan's writing on this subject, and the 'troublesome paradox' Kerlinger identified in that text received no further attention—not even from Kerlinger himself in his 1986 research methods text (Kerlinger, 1986). On the contrary, Kerlinger seems to have been quite content to leave the paradox unresolved, perhaps in confidence that it one day would be, and to continue to instruct additional cohorts of psychological researchers in the by-then widely accepted principles of hypothesis testing through aggregate-level statistical procedures.

Soon after the publication of Kerlinger's, 1979 text, I myself published an article in the *American Psychologist* in which I explained how the failure to respect the aggregate-general distinction was muddling theoretical discourse among psychologists concerned with the relationship between personality traits and behavior (Lamiell, 1981). Although the article received appreciable attention, neither it nor subsequent publications reiterating and elaborating on its central point, which was, again, fundamentally consistent with Bakan's (1955, 1966) advisories (cf. Lamiell, 2019), have significantly altered the research methods and interpretive practices within 'psychology's' mainstream. On the contrary, and notwithstanding the fact that the critical appraisal of those methods and interpretive practices set forth in the works just mentioned has never been successfully refuted, the critiqued methods and interpretive practices continue to this day (Harré, 2006). It is on this basis that I have branded mainstream thinking in 'psychology' *incorrigible* (Lamiell, 2017). This incorrigibility is going to have to be outgrown if there is to be any hope for genuinely psychological studies to re-gain the territory that statism forced psychologists to surrender to psycho-demographers.

A Confusion within a Confusion

A major component of that incorrigibility has been and continues to be the stubborn persistence of a false understanding of a kind of knowledge for which the German philosopher Wilhelm Windelband (1848–1915) invented the term *nomothetic* (Windelband, 1894/1998). Windelband's neologism was grounded in the ancient Greek *nomos,* meaning 'law.' Nomothetic knowledge is thus an assertion (thesis) of a law or lawful regularity. It captures what is recurrent across multiple instances of some phenomenon. In Windelband's terms, such knowledge reveals 'what always is' *(was immer ist)* the case when the phenomenon in question recurs (Windelband, 1894/1998). Windelband argued that the pursuit of nomothetic knowledge is the overarching knowledge objective of the natural sciences *(die Naturwissenschaften),* with physics being the most prominent among them.

Windelband's major concern in the 1894 work just cited was to counter what he saw as an ascendant over-valuation of the natural sciences and 'nomothetic' knowledge, and a commensurate de-valuation of the human sciences (or humanities), i.e., *die Geisteswissenschaften*—history being the most prominent among them.[6] In the human sciences, the primary knowledge objective is a nuanced account of historically situated—and quite possibly unique—phenomena, i.e., of the temporal and substantive contours of particular happenings or doings. Such knowledge Windelband labeled with the neologism *idiographic.* The first portion of that word 'idio-' is identical to that of the more widely familiar word 'idiosyncrasy,' and, indeed, both terms point to the particularities of individual cases or happenings. Such particularities might very well be unique, and knowledge of those particularities would therefore capture 'what once was' *(was einmal war).* Such knowledge is valuable, Windelband argued, and would be left completely out of consideration if one's quest were always and exclusively to discover what is nomothetic, or *common to all* of the individual cases under considera- tion, and therefore 'general' in the sense of the term understood by the late 19th century pioneers of experimental psychology (refer to discus- sion of this point in Chapter 1). Windelband argued that the intellectual health of the empirical sciences *(Erfahrungswissenschaften)*, which in the late 19th century included *both* the natural sciences *and* the human sciences, would be maximized by respect and appreciation for *both* 'nomothetic' *and* 'idiographic' knowledge. We will revisit this point at the conclusion of this chapter.

The failure to distinguish general type propositions from aggregate type propositions within the mainstream of 'psychology,' a failure that had become widespread within the discipline by the mid-20th century (hence the need seen by Bakan to address the matter; refer above), nurtured the mistaken belief that the aggregate statistical methods proper to population-level investigations were (and are) suited to the discovery of general lawfulness in—hence *nomothetic* knowledge about—the psychological doings of individuals. After all, so it was thought, since studies of populations entail observations of many individuals, the knowledge uncovered through the statistical analysis of those many observations must be interpretable as knowledge about each of the many individuals on whom those observations had been made. It would therefore be appropriate to regard that knowledge as knowledge of the studied individuals 'in general,' thus qualifying that knowledge as 'nomothetic' in just the sense intended by Windelband.

Long overlaying this utterly mistaken thinking was and remains the prevailing determination among mainstream psychologists for their field to be—and to be recognized as being—on a par with the natural sciences (refer to discussion of Ebbinghaus in Chapter 2).[7] This desire has undoubtedly redoubled mainstream thinkers' determination to regard statistical investigations of variables marking differences between individuals and groups—differences which, it must be remembered, can sensibly be defined only for aggregates of individuals—as suited to the quest for nomothetic knowledge about the psychological doings of individuals, and therefore altogether consistent with the insistence on prosecuting psychology as a natural science.

The conceptual problem here is that while statistical investigations of variables marking differences between individuals or groups could, in principle, yield nomothetic knowledge *of populations*, i.e., knowledge of 'what always is' when the variables under consideration are examined statistically in population after population, that knowledge would *not* qualify as nomothetic knowledge of the individuals *within* the investigated populations, and it is the latter sort of knowledge that is sought within psychology.

The confusions here are deep and abiding in the thinking of contemporary mainstream 'psychologists,' and, as history has shown, that thinking is firmly resistant to the changes demanded by close and critical conceptual analysis.[8] Again, however: if a genuine psychology is ever to regain the intellectual territory it long ago ceded to psycho-demography by paradigmatically adopting population-level methods of investigation, those confusions are going to have to be acknowledged and eliminated.

If and when such developments come to pass, the question left begging will be: How ought the 21st-century revival of psychological science proceed? It is this question to which the remainder of the present chapter is addressed.

Reviving Psychological Science

In the foregoing discussion, reference has repeatedly been made to the fact that the ideas and investigative practices that have undermined psychological science and replaced it with psycho-demography have dominated 'within the mainstream.' The reason for this emphasis is that not all within the field have adhered to the discipline's doctrinaire research methods canon. On the contrary, there has long been and continues to be a relatively small but significant minority of thinkers that has deliberately distanced itself from the majority's investigative practices in favor of methods that are actually suited to the objective of advancing our understanding of the psychological doings of persons. In concluding this chapter, I will mention briefly a few of these initiatives—far fewer than could and would be mentioned did available space permit—and will advocate the return to the conception of psychological science held by Windelband when, as discussed above, he introduced the distinction between 'nomothetic' and 'idiographic' knowledge nearly 130 years ago.

Restructuring Psychological Experimentation

The first point to be stressed here is that there is neither need nor intention to exclude from a revived psychological science investigative methods that are experimental and quantitative in nature. As I have done before (cf. Lamiell, 2003, pp. 246–263), I would invite readers with concerns on this point to consider studies that I directed many years ago to test the proposition that lay persons' judgments of the personality characteristics of oneself and others are made not through cognitive comparisons of the targets with the memory traces of judgments previously made of other targets (empiricist thinking), but rather through an implicit consideration of what is-rather-than-is-not the case about the particular target being judged in a specific instance (rationalist thinking). Mathematical models formally representing particular versions of these two different judgment processes were formulated, and applied to arrays of information about the behavioral patterns of each of a number of target individuals in order to mathematically generate alternative predictions of where the subject would position each of those targets along designated personality trait rating scales. For each subject, each of the two alternative sets of predicted ratings of each target was evaluated for its degree of correspondence to the actual ratings of that target made by that subject. This made it possible to index quantitatively for each subject, and on a target-by-target basis, the power of each of the two mathematical models of the judgment process for predicting that subject's ratings of the targets.[9]

For the record, data analysis pointed decisively to the superiority of the rationalist model of the judgment process over the empiricist model, and while for a handful of cases, analysis revealed no clear superiority of either model over the other, there were no cases in which the empiricist model proved superior to the rationalist model. These findings thus established the extensive—though not exceptionless—*generality* of the rationalist model as a formal representation of the psychological process governing subjective judgments of one's own and others' personality characteristics.

Within the framework of these investigations, as should be abundantly clear, 'generality' refers to the regularity with which the results obtained with one subject were replicated in the results obtained with other subjects. 'Generality' does *not* refer here to some singular experimental result defined for the collective of investigated subjects as a whole, or to the outcome of a statistical test comparing some sort of average values obtained for two or more sub-groups of subjects. On the contrary, in the specific case of the research reported by Lamiell and Durbeck (1987), fully 68 subjects were investigated, with the study of each subject being an experiment unto itself. In this epistemically crucial design feature, then, the work by Lamiell and Durbeck (1987) mimicked the design of the original experimental psychology experiments of the late 19th century (refer to Chapter 1), and bore no similarity at all to the psycho-demographic studies that came to the fore during the 20th century. Precisely for that reason, the work shed light on a psychological doing of individuals—namely, the process of judging one's own and others' personality characteristics—in a way that psycho-demography cannot.

While the nature of the judgment process through which subjective personality judgments are formulated can be seen to have important theoretical implications (see Rychlak, 1988), that is not the issue with which I am primarily concerned here. Rather, my purpose is simply to point interested readers to one concrete example of how experimental inquiry can proceed within a revived psychological science in a way that permits rigorous empirical tests *at the level of the individual* of a hypothesis concerning a psychological process that is understood theoretically to be transpiring *at the level of the individual.* Here, the level of theoretical concern, on the one hand, and the level of empirical investigation, on the other, are *matched.* Such matching cannot be achieved in a 'psychology' that continues to theorize at the individual level but conducts empirical analyses at the aggregate level. This is why Kerlinger (1979) was unable to solve the 'troublesome paradox' to which he himself had pointed (refer above), and why no one else has been (or ever will be) able to accomplish that feat, either. If a psychological science would enlighten us concerning the doings of individuals, then psychological scientists must study individuals. There is no alternative.

The truth of this last point has been fully respected in the development over the past two decades of a framework for psychological experimentation called Observation-Oriented Modeling (OOM) (Grice, 2011, 2015; Grice, Huntjens, & Johnson, 2020). Elsewhere, I have given a concise (and hence simplified) illustration of how OOM works in the conduct of psychological experimentation (see Lamiell, 2019, pp. 156–161), and so will not repeat that discussion here. Instead, and while urging readers to consult the above-cited works by Grice and colleagues (as well as others left un-cited here), I will reiterate three features of OOM that are of most fundamental conceptual relevance to our present concerns:

1 The use of OOM requires the investigator to specify as precisely as possible the theoretically anticipated experimental outcome for each individual research subject. This requires *theoretical clarity,* and thus discourages the 'dustbowl empiricism' that has so long fed the *scientistic* 'psychology' described earlier.
2 On the basis of (1), OOM proceeds in a way that enables an investigator to empirically examine *for each individual research subject* the extent to which the theoretically expected outcome for that subject is matched by the actually obtained empirical outcome for that subject.
3 On the basis of (2), it is possible in OOM to index the *generality* of experimental findings across individual research subjects, where, just as in the case of the research by Lamiell and Durbeck (1987) discussed above, 'generality' refers to the regularity with which theoretical expectations are confirmed by individual-level empirical findings, and not to some aggregate-level empirical outcome defined only for the collective of subjects or for sub-groupings thereof.

In a psychological science revived in the 21st century, high-quality experimental inquiry will consistently satisfy all three of these *desiderata,* something that cannot be accomplished by means of psycho-demographic methods.

Expanding the Space for Qualitative Investigations

As significant a role as systematic experimental inquiry could and should have in a revived psychological science, it will not and ought not to be the only mode of inquiry in any version of that revival that would qualify as critically personalistic. On the contrary, William Stern himself explicitly advocated over the entire course of his career the use of qualitative methods in the study of persons apart from—and beyond—what could be achieved through experimentation and the use of standardized psychological tests. For example, referring in the passage quoted below to qualitative methods as 'non-experimental' methods of investigation, he argued as follows in 1911:

> (By no means do systematic testing and experimentation) render non-experimental (i.e., qualitative) methods of investigation superfluous. To be sure, tests (and experiments) can supplement such methods. But tests (and experiments) are also supplemented by such methods, are dependent upon such methods for the confirmation and elaboration

of what they reveal, and in many cases must give way to what is revealed by those methods. (Stern, 1911, p. 105–106, parentheses added)

One highly promising contemporary initiative along the lines advocated by Stern as alternative to testing and experimentation is that of *narrative* psychology. In an Oxford University Press book series under the editorship of the accomplished and highly regarded Mark Freeman, psychologist Brian Schiff issued in a 2017 contribution to Freeman's series a fervent plaidoyer for a turn in psychological studies away from variable-centered inquiry, which has given us what I have termed psycho-demography, and toward an approach in which the accounts persons give of their own lives, or segments thereof, are studied both for their content and for their structure. Schiff (2017) argues persuasively that narration is not only a method for conducting psychological research, but is also itself a mode of psychological functioning. He views the telling of stories as a medium through which individuals construct and re-construct their identities as persons (see also Schiff, 2019).

Among the qualitative methods specifically advocated by Stern was *biography*. He even devoted himself in a sub-section of the 1911 book cited above to a discussion of how a psycho-biographical work should be structured (see Lamiell, 2021, chapter 6 pp. 122–128). The contemporary psychologist Jack Martin (b. 1950) is one prominent advocate of psycho-biographical inquiry as an investigative method for psychological studies. Despairing of mainstream psychology's inability to advance our understanding of persons, Martin (2020) writes of gaining considerable satisfaction in that regard by turning to psycho-biography. More specifically, Martin has found psycho-biographical work

… especially applicable to the illumination of two pivotal matters in the psychology of personhood: (1) the developmental emergence of social psychological aspects of persons (e.g., self and other understanding, social and personal identity, perspective taking, moral and rational agency, and character and comportment) and (2) the study of individual lives in ways that might warrant provisional, yet productive speculation concerning persons more generally. Each psycho-biographical study constitutes a possibility for human being that potentially enriches our sense of available options for living satisfying and productive lives. (Martin, 2020, p. 113)

Both of the scholars whose qualitative research I have chosen to highlight here (and they are but two from among many others who might alternatively have been cited; cf. Schiff, 2019) made mention in the cited works of being criticized for advocating approaches to psychological studies that are non- or even anti-scientific. Some or another form of such criticism has been leveled at psychologists at least since the time that Windelband espoused the need for 'idiographic' knowledge within psychological science alongside 'nomothetic' knowledge (refer to discussion of this distinction above). Accordingly, it is on this recurrent charge of non- or anti-scientism that I will comment in concluding this chapter.

Reviving a Broad Conception of 'Science' as 'the Making of Knowledge'

Considerable headway can be made toward de-fusing the non-/anti-science charge by re-embracing the understanding of 'science' built into the German word for the term: *Wissenschaft.* The first segment of that word, *Wissen,* means 'knowledge.'[10] The second segment of the word, '-*schaft*,' derives from the verb *schaffen,* meaning 'to do' or 'to make' or 'to accomplish.' Thus is 'science' to be understood as 'the doing or making of knowledge.'

Some knowledge is of a purely *rational* sort, having no specific empirical content, and Windelband (1894/1998) referred to disciplines devoted to making such knowledge the 'rational' sciences *(die rationalen Wissenschaften):* logic and mathematics. Sciences with empirical content are referred to as the *Erfahrungswissenschaften,* the empirical sciences, and include *both* the *Naturwissenschaften* or 'natural' sciences *and* the *Geisteswissenschaften* or sciences of the mind or spirit or psyche. These are the 'human' sciences or, as they are often referred to in English, the 'humanities.'

Windelband clearly regarded psychology as a 'hybrid' discipline (my term), viewing its subject matter as falling within the domain of the human sciences, but its then-favored method of investigation as placing it within the domain of the natural sciences.[11] Psychology would therefore be understood properly as a discipline incorporating both 'nomothetic' knowledge and 'idiographic' knowledge. The point warranting particular emphasis here is that *both* in its prosecution as a natural science *and* in its prosecution as a human science (or 'humanity') psychology would be a 'Wissenschaft,' i.e., a discipline devoted to the making of knowledge. Only by arrogating the term 'science' for reference to *natural* science exclusively can psychology pursued as a humanity be dismissed as non-/anti-scientific.[12] Shorn of such dilettantism, psychological science revived in the spirit envisioned here will be able to flourish harmonically both as a natural science and as a human science. It should never have become otherwise.

Notes

1 It was for reference to such practical applications of psychological concepts and methods in domains outside of the discipline of psychology itself that Stern proposed the term 'psychotechnics' (Stern, 1903).

2 My designation of this term is not entirely fanciful. According to the contemporary historian, Theodore M. Porter (b. 1953), 'the term "statistics" derives from a German term, *Statistik,* first used as a substantive by the Göttingen professor Gottfried Achenwald [1719–1772] in 1749 ... . [While] its etymology was ambiguous and its definition remained a matter of debate for more than a century, ... most writers of the early nineteenth century agreed that it was intrinsically a science concerned with states [the German word for "state" being *Staat*], or at least with those matters that ought to be known to the "statist"' (Porter, 1986, pp. 23–24, brackets added).

3 Stern was able to finish writing this work only after he had fled from Nazi Germany to the Netherlands in 1934. A general psychology textbook, it was published in German, albeit by a Dutch publishing house (Nijhoff), the following year.

4 To illustrate this latter possibility, suppose that the latest nationwide census in the U.S. has shown that, on average, American families have 2.2 children. This would be an aggregate-level truth that would not—indeed *could* not—be true of *any* one family within the aggregate.

5 This notion would have astounded Wundt and Ebbinghaus and the other pioneers of psychology as an experimental science, a reality vividly reflecting how far contemporary 'psychology' has drifted from its conceptual origins.

6 Hence the title (in translation) of Windelband's text: 'History and Natural Science.'

7 This desire has sometimes been playfully but instructively referred to as 'physics envy.'

8 Within days of the publication in the *American Psychologist* of my 1981 article (refer to the discussion above), a senior colleague of mine congratulated me on that accomplishment, but commented with obvious disdain that the work was 'merely conceptual.' Better, I was invited to believe, would have been experimental proof of the validity of my argument, as if the points it raised were matters to be decided empirically rather than through careful critical examination for conceptual coherence! Thus transpired my own first professional collision with scientism.

9 For a fully detailed exposition of one of the studies being summarized here, see Lamiell and Durbeck (1987).

10 In its verb form, '*wissen*' means 'to know.'

11 It must be remembered here that the psychology on which Windelband was commenting was the experimental psychology being prosecuted by Wundt, Ebbinghaus, and the other pioneers of the discipline as an empirical science.

12 For one especially egregious example of this, consider the declaration by J. C. Nunnally that 'idiography is an anti-science point of view: it discourages the search for general laws and instead encourages the description of particular phenomena (people)' (Nunnally, 1967, p. 472, parentheses in original).

References

Allport, G. W. (1938). William Stern: 1871–1938. *The American Journal of Psychology, 51,* 770–773.

Bakan, D. (1955). The general and the aggregate: A methodological distinction. *Perceptual and Motor Skills, 5*, 211–212.

Bakan, D. (1966). The test of significance in psychological research. *Psychological Bulletin, 66*, 423–437.

Bennett, M. R., & Hacker, P. M. S. (2003). *Philosophical foundations of neuroscience.* Oxford, UK: Blackwell Publishing.

Gantt, E., and Williams, R. N. (Eds.) (2018). *On hijacking science: Exploring the nature and consequences of overreach in psychology.* New York: Routledge.

Grice, J. W. (2011). *Observation Oriented Modeling: Analysis of cause in the behavioral sciences.* New York: Academic Press.

Grice, J. W. (2015). From means and variances to patterns and persons. *Frontiers in Psychology, 6*, 1–12.

Grice, J. W., Huntjens, R., & Johnson, H. (2020). Persistent disregard for the inadequacies of null hypothesis significance testing and the viable alternative of observation-oriented modeling. In Lamiell, J. T., & Slaney, K. L. (Eds.), *Problematic research practices and inertia in scientific psychology* (pp. 55–69).

Harré, R. (2006). *Key thinkers in psychology.* Thousand Oaks, CA: Sage Publications.

Kerlinger, F. N. (1979). *Behavioral research: A conceptual approach.* New York: Holt, Rinehart, & Winston.

Kerlinger, F. N. (1986). *Foundations of behavioral research,* 3rd edition. Chicago: Holt, Rinehart, & Winston.

Kerlinger, F. N., & Pedhazur, E. J. (1974). *Multiple regression in behavioral research.* New York: Holt, Rinehart, & Winston.

Lamiell, J. T. (1981). Toward an idiothetic psychology of personality. *American Psychologist, 36*, 276–289.

Lamiell, J. T. (2003). *Beyond individual and group differences: Human individuality, scientific psychology, and William Stern's critical personalism.* Thousand Oaks, CA: Sage Publications.

Lamiell, J. T. (2017). The incorrigible science. In Macdonald, H., Goodman, D., & Becker, B. (Eds.), *Dialogues at the edge of American psychological discourse* (pp. 211–244). London, UK: Palgrave-Macmillan.

Lamiell, J. T. (2019). *Psychology's misuse of statistics and persistent dismissal of its critics.* London, UK: Palgrave-Macmillan.

Lamiell, J. T. (2021). *Uncovering critical personalism: Readings from William Stern's contributions to scientific psychology.* London, UK: Palgrave-Macmillan.

Lamiell, J. T., & Durbeck, P. (1987). Whence cognitive prototypes in impression formation? Some empirical evidence for dialectical reasoning as a generative process. *Journal of Mind and Behavior, 8*, 223–244.

Lück, H. E., & Löwisch, D.-J., (Eds.) (1994). *Der Briefwechsel zwischen William Stern und Jonas Cohn: Dokumente einer Freundschaft zwischen zwei Wissenschaftlern* (Correspondence between William Stern and Jonas Cohn: Documents of a friendship of two scientists). Frankfurt am Main: Verlag Peter Lang.

Machado, A., & Silva, F. J. (2007). Toward a richer view of the scientific method: The role of conceptual analysis. American Psychologist, *62*, 671–681.

Martin, J. (2020). Psychology's struggle with understanding persons. In J. T. Lamiell & K. L. Slaney (Eds.), *Problematic research practices and inertia in scientific psychology* (p. 102–115). London: Routledge.

Nunnally, J. C. (1967). *Psychometric theory*. New York: McGraw-Hill.

Rychlak, J. F. (1988). *The psychology of rigorous humanism* (second edition). New York: New York University Press.

Schiff, B. (2017). *A new narrative for psychology*. New York: Oxford University Press.

Schiff, B. (Ed.) (2019). *Situating qualitative methods in psychological science*. New York: Routledge.

Stern, W. (1903). Angewandte Psychologie [Applied psychology]. *Beiträge zur Psychologie der Aussage*, *1*, 4–45.

Stern, W. (1906). *Person und Sache: System der philosophischen Weltanschauung, erster Band: Ableitung und Grundlehre* [Person and thing: A systematic philosophical worldview, Volume 1: Rationale and basic tenets]. Leipzig: Barth.

Stern, W. (1911). *Die Differentielle Psychologie in ihren methodischen Grundlagen*. Leipzig: Barth.

Stern, W. (1935). *Die allgemeine Psychologie auf personalistischer Grundlage [General psychology from a personalistic standpoint]*. Den Haag: Nijhoff.

Stern, W. (1927). Selbstdarstellung [Self portrait]. In R. Schmidt (Ed.), *Philosophie der Gegenwart in Selbstdarstellung* (Vol. 6, ppp. 128–184). Leipzig: Barth.

Watson, J. B. (1928). *The ways of behaviorism*. New York, NY: Harper and Brothers.

Windelband, W. (1894/1998). History and natural science (J. T. Lamiell, Trans.). *Theory and Psychology, 8*, 6–22.

Wundt, W. (1913/2013). Psychology's struggle for existence (J. T. Lamiell, Transl.) *History of Psychology, 16*, 197–211. 10.1037/a0032319.

Part II

Toward a Critical Inter-personalism in the Grounding of a Socio-Cultural Ethos

4   Echoes of William Stern's Socio-Cultural Voice

As I noted in the preface to this work, William Stern made explicit his understanding that the worldview he called 'critical personalism' offered not only a philosophical basis for psychological theorizing but also a framework 'for the grounding of cultural life' (Stern, 1923, p. 270). To the best of my knowledge, Stern never systematically elaborated his understanding of just how critically personalistic thinking should 'ground cultural life.' However, in various works that he published over the course of his scholarly life, he wrote in ways that vividly reflected his thinking in this regard.

The present chapter is devoted to a discussion of some of those works. My purpose here is twofold. First, I hope to awaken in readers some historical appreciation for William Stern's sense of the relevance of his scholarly efforts as a philosopher and psychological scientist to socio-cultural issues that arose in his time. Secondly, but of perhaps greater importance in the long run, I want to prompt in readers a sense of the possibilities of critically personalistic thinking for addressing contemporary issues of socio-cultural consequence. Realizing these two objectives should facilitate the work of the next two chapters, where we will consider some of the ways in which critically personalistic thinking could be brought to bear on discourse concerning the issue of racism in American society (Chapter 5), and on other domains of interpersonal life (Chapter 6).

On the Ethical Significance of Tolerance

Perhaps Stern's earliest scholarly foray into the domain of the socio-cultural was a lecture he gave on March 15, 1900, when he was not yet 30 years old. The lecture, the text of which, to the best of the present author's knowledge, was never published, was given at a meeting of an organization known as the 'Society of Brothers,' and was titled (in translation) 'On the Ethical Significance of Tolerance' *(Über die ethische Bedeutung der Toleranz)* (Stern, 1900a).[1]

DOI: 10.4324/9781003375166-6

Stern opened his remarks by posing the question 'What is tolerance?' using, as he did in the title of his presentation, the German word *Toleranz*. He then pointed out that the word had a variety of connotations captured by two related but nevertheless distinct meanings that he took to be implied by the German cognates *Duldung* and *Duldsamkeit.*

Duldung, Stern explained, referred to 'behavior that one practices, ... [it is] a practical stance adopted by an individual or a collective, e.g., the state, with respect to the representatives of a view that does not correspond to the individual's or collective's own.' In my judgment, the meaning Stern attributed to *Duldung* is well captured in English by the term *sufferance*. It is a contingent form of tolerance that is ultimately a matter of expediency. Stern wrote that, historically, this variety of tolerance emerged at the time of the Roman Empire with '... a realization that a multiplicity of views, customs, and beliefs could exist side-by-side ... . The tolerance of this period was not a truly moral virtue, but rather a kind of generous big-heartedness, bound to a certain form of statecraft *[Staatsklugheit].*'

One of the limitations of this form of 'tolerance' is that, as a relatively superficial form of behavior, it does not entail earnest critical engagement with others who are different from oneself. Hence, it can easily degenerate into *indolence,* or, in the contemporary English vernacular of those of a certain age, an attitude of 'whatever.' Indolence, Stern argued, is the true and equally non-virtuous opposite of intolerance. He was fully cognizant of the dangers entailed by an inclination to accept anything and everything with an indifferent shrug of the shoulders, and he warned that 'the damages that can be done by hateful intolerance are not greater than those which can result from indolence.' He underscored this belief as follows:

> Fortunate is one who can still get worked up and become outraged, ... but precisely where one has such feelings, one cannot be tolerant ... . If one proclaims tolerance at any price a virtue, then one would have to abandon every evaluation, every judgment, and every criticism ... . [Were we to do that], we would surrender the best aspect of our human nature and of human ability: the ability to *evaluate.*

As noted in Chapter 2, the ability to evaluate was, for Stern, a critical difference between *persons* and mere *things:* persons are, by their very nature, actively evaluative beings, whereas things are inherently passive entities that can only be evaluated. It was in just this realm of evaluation where Stern located *Duldsamkeit*, or genuinely virtuous tolerance. Unlike *Duldung,* that form of behavior—or posturing—guided by considerations of expediency, Stern understood *Duldsamkeit* to be

... a sentiment that one has ... . [It is] an aspect of character relating to our judgments, ... making it possible for us to ... find value in and justification for an opinion or perspective that deviates from our own.

Stern's was thus a decidedly and quite deliberately Aristotelian conception of tolerance as a virtue of character situated between the opposites of intolerance, a vice that erupts when contingencies no longer recommend sufferance, and indolence, a vicious form of lazy acquiescence.

One especially instructive feature of Stern's 1900 presentation is his discussion of historic developments in the treatment of Jews. Painting with a very broad brush, he argued that before the 18th century, such 'tolerance' as was accorded Jews at all was in the nature of *Duldung,* i.e., sufferance. The acceptance of Jews in the towns and villages rested on the belief that 'in certain areas of life, such as trade and business, Jews were indispensable.' The acceptance of Jews, in other words, was contingent upon their economic serviceability.

However, as the 18th century gave way to the 19th, Stern argued, 'every movement that carried the Jew toward emancipation and civil rights emerged from a sentiment of tolerance which, as a characteristic of the larger people, was a product of the period of Enlightenment.' So, by 1900, in Stern's' view, that comparatively superficial form of tolerance he saw connoted by *Duldung,* i.e., what I am here calling 'sufferance,' had been replaced, at least where the treatment of Jews was concerned, by that enlightened and genuinely virtuous form of tolerance he understood to be connoted by *Duldsamkeit.*

It is sobering—but very instructive—to realize that the author of the view being presented here was himself a Jew, and that three decades after his 1900 speech on the ethical significance of tolerance he had to flee his native Germany in order to escape Nazi persecution. That Stern could have misread so egregiously the status of Jews in German society at the turn of the 20th century can be seen to underscore the importance of the distinction he was drawing: *Duldung,* or contingent sufferance, can easily masquerade as *Duldsamkeit,* i.e., genuinely virtuous tolerance. The fact of this matter warrants our serious reflection today, in the face of the enormous challenges now confronting us in our multi-cultural world.

Echoes of Stern's Socio-Cultural Voice in the Domain of Child Psychology

A scant three weeks after he delivered the lecture just discussed, the young William Stern and his wife of little more than a year, Clara, welcomed into their world daughter Hilde, who was born on April 7, 1900. William Stern recorded his observations of the event in writing, and in so doing authored the first entry in what would prove to be an 18-year-long diary project culminating in upwards of 5,000 handwritten pages of observations of each of the three children that Clara and William Stern would parent. Following Hilde, son Günther was born in 1902, and second daughter Eva arrived in 1904. For each child, the diary record began at birth and lasted until that child reached puberty (Stern & Stern, 1918).

The accumulating data in these diaries would provide the bulk of the empirical basis for several of Stern's publications in the sub-field of what was then referred to as 'child psychology.' One of those is a monograph that Stern co-authored with Clara and published in 1909 under the title (in translation) *Recollection, Testimony, and Lying in Early Childhood*.[2] That work merits some discussion here because, in their treatment of lying by children, the Sterns made clear their understanding of the matter as not only an aspect of children's psychological development but also a matter of substantial socio-cultural significance.

On the Development of Lying in Children

It was in the penultimate chapter of the 1909 monograph where Sterns addressed themselves specifically to the origins of lying, and to the measures that parents and other caregivers could take in an effort to prevent habitual lying from developing. They began that chapter by observing:

> Different perspectives on the moral aspects of early childhood are nowhere in greater conflict than with respect to the pedagogical problem presented by lying.
>
> On the one hand, the belief prevails that the small child is by nature not merely amoral but indeed anti-moral, and that, therefore, lying is among the earliest manifestations of an egoism that has yet to be brought under control. On the other hand, there is the conviction that the naiveté of the child entails an innocence and lack of guilt, and that therefore the caretakers alone must bear the responsibility for the child's lies. (C. Stern & W. Stern, 1999, p. 129)

Here the Sterns quoted in the original French the Swiss philosopher Jean Jacques Rousseau (1712–1778): Rendered in English, the quote states that 'everything is good when it leaves the hand of the Creator, but degenerates in the hands of man' (cf. C. Stern & W. Stern, 1999, p. 129).

The Sterns named the two then-ascendant perspectives on the development of lying 'nativistic' and 'empiricistic,' respectively, and then proceeded to defend the critically personalistic view that 'just like every other psychological function, lying must be seen as a consequence of the convergence of inner and outer factors' (C. Stern & W. Stern, 1999, p. 129; refer to discussion of the concept of 'convergence' in Chapter 2).

The Sterns explicitly branded as erroneous the nativistic notion that lying *per se* is a natural tendency that children have. They argued instead that certain natural tendencies that children do have can converge with certain outer influences to give rise to lying. For example, children are naturally inclined to defend themselves against the threat of pain or physical danger, and in convergence with a social milieu dominated by adults' ongoing use of corporal punishment, the child may defensively resort to telling lies. Thus can the habit of lying take root.

Children also have, according to Sterns, a natural tendency to imitate, and when they are very young this imitation can be indiscriminate. In a milieu where adults themselves tell what are often casually dismissed as mere 'little white' lies, or where the child is even recruited into the telling of such lies, e.g., by following mother's orders to tell a caller at the door that 'Mother is not home' when she does not wish to be visited, habitual lying can begin to develop in the child.

These and other examples of person-world convergence that the Sterns discussed in this context led them to address the question of how best to work toward the prevention of lying from ever developing in a child to begin with. From an extreme nativistic viewpoint, efforts in this direction would, by definition, be futile. On that view, the child is bound to eventually lie, period.

At the same time, Sterns argued that the aforementioned Rousseau had gone too far in the empiricistic direction in his *Emile* when he argued that one should not demand the truth from a child and, in the process, induce the child to conceal the truth. In this context, they opposed Rousseau's declaration that 'the more the child's well-being is made independent, be it of the will of others or of their judgment, the less will be the child's interest in lying.'

Rousseau's philosophy of childrearing was a highly individualistic one, and Sterns' commentary on Rousseau further reflects the non-individualistic nature of the critically personalistic framework within which they were working. They commented on the above-noted counsel by Rousseau as follows:

One can well understand Rousseau's pleas as a reaction to prevailing, overly strict child-rearing practices. But to the extent that he is struggling against excessive discipline, he also undermines, to a considerable extent, self-discipline, and it is precisely those child-rearing practices that foster self-discipline that offer the best possibility of recruiting the child himself to the collective struggle against lying ... . A child whose parents teach her about the importance of maintaining self-control in general, a child who has learned to curb his own anger, or to forego a pleasure out of consideration for others, or to tolerate an unfairness—yes, a child who can take satisfaction in having achieved self-control—will overcome her own inclinations to lie. (C. Stern & W. Stern, 1999, p. 137)

What warrants emphasis in all of this is the Sterns' conviction as psychologists that while the social milieu in which a child is raised is certainly a critical factor in that child's development, it is not a causal force producing certain behavioral outcomes in a strictly mechanistic fashion the details of which we have only to discover through empirical research. Rather, the social milieu is an 'around-world' (an *Umwelt*) that sets parameters within which the child will ultimately act *deliberately*, in accordance not only with his natural inclinations but also in consideration of societal ('heterotelic') *values* that he will provisionally either embrace or reject, and for which he will ultimately be accountable. Also clear in all of this is the Sterns' understanding that lying is not just an individual/psychological phenomenon, but also a socio-cultural phenomenon of great consequence for the very fabric of the larger community. Hence, it is a phenomenon for which parents, educators, and adult caregivers all share moral responsibility.

On the Identification of Highly Talented Pupils

Another context in which the socio-cultural voice of William Stern the child psychologist found expression was provided by his work on the identification of pupils who would be characterized today as 'gifted and talented.' Work bearing on this subject formed a major part of Stern's research program in Hamburg beginning in 1916, when he moved there to succeed Ernst Meumann (1862–1915) as Director of the psychological laboratory that Meumann had founded in 1911. Stern's work on this subject continued through the establishment, due in part to his efforts, of the University of Hamburg in 1919, and on until 1933, when the accession to power in Germany of the Nazis abruptly ended all of his university-related activities.

In a letter to his friend and colleague, the Freiburg philosopher Jonas Cohn (1869–1947), written in October of 1918, Stern wrote that

... the problem of selecting and advancing talented young persons is everywhere in need of discussion. In fact, the questions here are not just in need of psychological investigation, but must also be considered in terms of their ethical, socio-political, and pedagogical facets. (Stern letter to Cohn, October 11, 1918; reprinted in Lück & Löwisch, 1994, p. 115)

A decade later, Stern would underscore this point in one of his publications on the topic:

> The insight that the advancement of highly capable youth would be a socio-ethical task of the first order has spread further and further in recent years ... . We stand before an 'ethics of ability,' such that, on the one hand, the people at large recognize their duties relative to those talents growing within our midst, and, on the other hand, that individuals blessed with a special ability not be permitted to see in it a private privilege which they enjoy, but a special duty to themselves and to the entire society. (Stern, as quoted in Feger, 1991, p. 98)

We see here once again the non-individualistic nature of critically personalistic thinking.

In her discussion of Stern's historical contribution to work in this area, Feger (1991) noted that following World War II, American researchers came to dominate this line of inquiry (as well as many others in psychology). In as much as the ethos in the U.S. was—and remains—predominantly individualistic, and in that respect incompatible with critical personalism, it is not surprising that Stern's influence in this domain (too) has not been lasting. Arguably, an infusion of critically personalistic thinking, not only into psychologists' work in this area, but also, by extension, into the perspective of educators in the primary and secondary school classrooms of today, could very well redound to the considerable benefit of our communities. In any case, this is certainly the way Stern thought things should be in the pedagogical communities of his time.

On the Practice of Psychoanalytic Psychotherapy with Children and Adolescents

For all of the foregoing, it seems safe to say that there was no issue that exercised the socio-ethical sensibilities of William Stern as a child psychologist more than the issue of the conduct of psychoanalytic psychotherapy with children and adolescents. Without question, his concerns in this matter issued from his respect for the importance of basing any and every intervention in people's lives on good scientific practices, and he had considerable doubts about the soundness in this regard of the work of Freud and his followers. Those doubts were given clear expression as early as 1901, when Stern published a largely unfavorable review of Freud's *The Interpretation of Dreams*, which had appeared the year before.

In his review, Stern did compliment Freud on, among a few other particulars, the boldness of Freud's effort to extend the understanding of dream life 'down into the core of the world of affects' (Stern, 1901, p. 131). However, Stern followed up his compliments with harsh criticism, most of which was aimed at Freud's method. After articulating several specific concerns in this regard, Stern wrote:

> The inadmissibility of [Freud's] approach to dream interpretation as scientific method must be emphasized, because the danger is great that uncritical souls could get comfortable with this engaging mind game, and that we, in consequence, would slip into an entirely mystical and arbitrary exercise, whereby one could then prove everything with anything. (Stern, 1901, p. 133)

Nor did Stern pass the opportunity to convey explicitly his doubts about the validity of Freud's theoretical convictions regarding the role of sexual themes in unconscious psychological life. On the contrary, he wrote:

> A particular tendency to impose a sexual interpretation on all possible and impossible dream contents plays such a prominent role in [Freud's] book that it would be pointless to provide an isolated example. (Stern, 1901, p. 133, brackets added)

Clearly, the concerns Stern was voicing here were primarily epistemic in nature, and these concerns continued to play a role when, more than a decade later, he challenged the legitimacy of practicing psychoanalytic psychotherapy with children and adolescents. But in this latter context, the stakes were even higher because ethical concerns entered in as well.

By the time Stern first made public his concerns about such practices, he and his wife Clara were some 13 years into the diary project that was mentioned earlier in this chapter. So, while deploring the Freudians' lack of persuasive empirical evidence to positively *support* their theoretical assumptions concerning the sexual nature of even young persons' unconscious lives, Stern also regarded himself as well-supplied with empirical material that strongly *contraindicated* the validity of those assumptions. If he was correct in this regard, then the Freudians were not only objectively incorrect in their theoretical assumptions about the nature of young persons' unconscious mental lives, but they were acting unethically by proceeding to intercede in the lives of young persons as if those theoretical assumptions had been scientifically validated. Hence the resolve with which he wrote in a letter to Jonas Cohn dated June 9, 1913:

> What they [the psychoanalysts] have done now exceeds all comprehension.

Psychoanalysis has become a pedagogical danger, and it is high time for the Commission for Youth Studies to stand up in opposition to this. (Stern letter to Cohn, June 9, 1913; reprinted in Lück & Löwisch, 1994, p. 91)

A movement in public opposition to the Freudians in this matter was in fact initiated by Stern and numerous colleagues during an October 1913 meeting in Breslau of the Society for School Reform, a subsection of the Commission for Youth Studies. At that meeting, the so-called Breslau Warning was issued, bearing the signatures of 31 members of the Society for School Reform, including William and Clara Stern. That document proclaimed the following:

The undersigned members of the section for Youth Studies within the Society for School Reform regard it as their duty to notify friends of youth and pedagogy of the dangers that exist due to attempts to apply psychoanalytic methods to children and adolescents ... . Without taking any position on the scientific significance of psychoanalytic theory and the therapeutic application of its methods with adults, the undersigned declare:

1 The claim that psychoanalytic methods prove that previously conducted research on children has been misdirected, and that only through psychoanalysis has a scientific child psychology become possible, is unjustified.
2 The extension of psychoanalytic methods to application as part of routine child-rearing practices is to be rejected. The reason for this is that psychoanalysis can lead to a lasting psychological infection of the treated with premature sexual fantasies and sexual feelings, and thus deprive children of their innocence in a way that presents a great danger for our youth. The various successes in child rearing that have been claimed by practitioners of psychoanalysis are greatly outweighed by the damages that will be done to the immature mind. (Reprinted in Graf-Nold, 1991, p. 69)

Not content with this warning, Stern would author a full-length article the following year, elaborating his concerns over the practice of psychoanalytic psychotherapy with children and youth. In that article, he responded to intimations that he and Clara had deliberately omitted material of a sexual nature from their publications. Stern wrote:

The fact that we have not written about sexual matters [in our works on child psychology] is not because we have concealed anything, but rather because nowhere in our intensive observations have we discovered the slightest trace of sexual undertones in our children's recollections. (Stern, 1914, p. 86)

It was largely on this basis that Stern went on to assert in the 1914 article that the practice of psychoanalytic psychotherapy with young persons was 'not only an egregious error (*Verirrung*) but a pedagogical sin (*Versündigung*)' (Stern, 1914, p. 91).

Nowhere in my readings of Stern's works have I found a passage more vehement than this one, and I think that the forcefulness of Stern's language here reflects the utmost seriousness with which he and Clara regarded the moral imperative of not depriving children prematurely of their innocence.

Echoes of Stern's Socio-Cultural Voice in the Domain of Psychological Testing

On Psychological Testing as a Socio-Cultural Issue

Reference was made early in this chapter to Stern's steadily growing disdain for developments within the psychological testing movement. As the founder of the sub-field of 'differential' psychology, within which the testing movement took root and has proliferated (Lamiell, 2003), Stern's concerns in this area surfaced time and again over the last 20 years or so of his professional life. Accordingly, it is to this topic that the discussion now returns.

It was in 1900 when Stern published the first systematic consideration of differential psychology as a sub-discipline of the field, distinct from but coordinated with the general experimental psychology of the time. The title of the book (in translation) was *On the Psychology of Individual Differences: Toward a 'Differential' Psychology* (Stern, 1900b). In that book, Stern identified two overarching scientific objectives that would be served by differential psychology: the first of these was the basic science objective of *Menschenkenntnis, i.e.,* 'knowing humans.' Stern was thus saying that differential psychology could advance our knowledge in this domain beyond what could be achieved through general experimental psychology alone. The second overarching scientific objective to be served by differential psychology, Stern (1900b) argued, was the practical or applied science objective of *Menschenbehandlung, i.e.,* the handling or treatment of humans. In this domain, Stern's vision was that the knowledge produced by differential psychology could enable a more salutary allocation of human resources, e.g., in the assignment of pupils to various courses of study in school, or in the deployment of adults in the workplace.

Both *Menschenkenntnis* and *Menschenbehandlung* would require for their realization the development and implementation of methods for diagnosing individualities that would be as accurate and comprehensive as possible. However, Stern had grave doubts that this objective could be achieved by relying solely on standardized tests. If he only hinted at those doubts in his 1900 book, he expressed them quite explicitly in the sequel to that book, a text that he published in 1911 titled (in translation) *Methodological Foundations of Differential Psychology* (Stern, 1911). His concerns in this regard would be further expressed in numerous other publications that would follow over the next two decades.

In chapter 6 of the 1911 book, a chapter that was devoted specifically to a discussion of testing as a form of systematic psychological research, Stern wrote:

> Of course, it is possible that studying an individual through the administration of a battery of tests can yield material that will be valuable for many purposes of comparative research. But for the specific goal of fashioning a comprehensive characterization of the psychological functioning of an individual, precious little *(verschwindend wenig)* is to be gained in this way. One can determine how the tested person behaves at a specific point in time with respect to ten or twenty functions, but this momentary behavior permits no conclusions whatsoever concerning any lasting trends in the tested functions. Even less is offered by way of knowledge about the characteristic functioning of the person in other areas, and it is precisely the most characteristic traits that lie furthest from the more peripheral performances examined in these tests. (Stern, 1911, p. 90)

With these and other considerations in mind, Stern concluded his chapter by exhorting his readers to keep firmly in mind that

> The test is only *a*—and not *the*—method for examining individuality. By no means does it render [other] methods of investigation superfluous. To be sure, tests can supplement such other methods. But tests are also supplemented by those other methods, are dependent upon those other methods for the confirmation and elaboration of what they reveal, and in many cases must give way to what is revealed by those other methods. Psychological testing *per se* is to be regarded as a ... stopgap measure *(Notbehelf)* when time constraints or other circumstances will not admit of supplemental methods. It also serves as a method of preliminary investigation for the purpose of selecting from a large group some particular individual as a subject of further and more detailed ... investigation. (Stern, 1911, p. 106)

From a strictly epistemic standpoint, i.e., when considered solely with regard to the basic knowledge objective of *Menschenkenntnis*—the understanding of persons—the problem that results from viewing persons only through the lenses provided by standardized tests is that each tested individual cannot be regarded as a distinctive individual, but must instead be viewed as one of indefinitely many instantiations of some subset of the psychometric categories defined by the test(s) in use. Viewed this way, each individual is, in principle, replaceable, both by and for, any and every one of the other individuals who instantiate the same subset of categories. But if, as Stern the critical personalist insisted, persons are unique, then portraying them in this way results in falsehoods, hardly an epistemic *desideratum* for a genuinely scientific psychology.[3] Thus did Stern insist at the Fourth International Congress for Psychotechnics, held in Paris in 1927 (the text of his presentation there would be published two years later):

> A person forms a unity and has depth. The human is not a mosaic, and therefore is not to be portrayed as a mosaic. All attempts to portray a person in terms of an array of test scores are fundamentally false. By dissecting the person in accordance with the elementary tests applied in isolation, we do not get closer to the essence of the personality. On the contrary, we move further away from it. (Stern, 1929, p. 65)

So much for strictly epistemic concerns. If we now shift our focus from the basic knowledge objective of *Menschenkenntnis* to the decidedly more applied knowledge objective of *Menschenbehandlung*, i.e., the treatment or deployment of persons as human resources, the problem is no longer strictly epistemic but is also socio-ethical. When psychometricians, including but not limited to those working within the then-nascent field of psychotechnics, not only *understand* the individuals who have been submitted to their psychological tests as mere instantiations of category clusters in the sense just described, but proceed to *treat* those individuals in accordance with the dictates of statistical evidence regarding the relevance of those category clusters to the ends being sought by the institutions, i.e., the schools, businesses, governments, etc. which have retained those very psychometricians (for a fee), real persons' lives are being affected in ways, and on grounds, which demand justification. Thus Stern argued at the just-mentioned 1927 Congress for Psychotechnics in Paris:

> In those companies relying on psychotechnical selection procedures, it must be remembered that one is not dealing with machines or materials, whose quality and economic significance for the company is in fact expressible through test scores, but rather with human beings, whose occupation is a part, and, indeed, a very essential part, of their entire lives. (Stern, 1929, p. 72)

Two years later, Stern reiterated these concerns, using rather more pointed language. At the Seventh International Conference for Psychotechnics, held in Moscow in 1931, he urged his colleagues to bear in mind that

the psycho-technician does not work with machines, or with wares, or, in short, with things, but rather with human beings. Under all conditions, however, human beings are and remain the centers of their own psychological life and their own worth. In other words, they remain persons even when they are studied and treated from an external perspective and with respect to others' goals. [And] if today the word 'psychotechnics' is sometimes uttered with disdain, that is due to the implicit or explicit belief that psycho-technicians not only intercede but interfere in the lives and rights of the individuals with whom they deal. The feeling is that psycho-technicians degrade persons by using them as means to others' ends. (Stern, 1933, p. 55)

Stern's remarks both at the 1927 Paris conference and at the 1931 Moscow meetings can be seen as entirely within the spirit of words he had written in a letter to Jonas Cohn more than a decade earlier, on October 5, 1916. Noting the rapid development of the sub-discipline of occupational psychology and its steady advancement as an applied science, Stern wrote:

I am anxious to see where these efforts will lead. For my part, I will strive for an approach that is not dominated by Taylorism,[4] but instead one in which social ethics plays the leading role in occupational aptitude testing. What we need is not an industrial psychology but a psychology of capabilities. (Stern letter to Jonas Cohn, October 5, 1916; reprinted in Lück & Löwisch, 1994, p. 103)

Socio-Cultural Concerns about the Proliferation of Psychological Testing

Between the 1927 Paris conference and the 1931 Moscow meetings, Stern traveled to the U.S. to attend the International Congress of Psychology in New Haven, Connecticut in 1929. After that Congress, Stern spent several weeks traveling in the U.S. to visit various universities, so as to gain a better sense for the themes that were dominating the work of academic psychologists in the U.S. at that time. Upon his return to Germany, he authored an article in which he wrote about the impressions he had gained during his travels. In that connection, he wrote:

The face of American psychology is characterized much less by laboratory experimentation than by testing procedures ... . Since the (First World) War, during which the entire American army was tested by means of a simple, standardized procedure for measuring intelligence, testing methods have been extended in ways that are astounding and almost troubling ... . Seventeen years ago, when I introduced the concept of the 'intelligence quotient' as a measurement principle for such intelligence tests, I had no idea that the 'IQ' would become a kind of world-wide formula and one of the most frequently encountered expressions in American technical jargon. But beyond that, batteries of tests for countless other psychological functions such as spatial perception, manual dexterity, attention, suggestibility, knowledge, arithmetic ability, character traits, etc. have now been developed, standardized, and put into use, always with emphasis on the objective, quantitative norm, with reference to which the single case is then compared. At times, the primary objective in America seems to be to exercise technique, to obtain numerical measures that can be correlated and statistically analyzed. [What is clear] in all of this is the danger of mechanization, and it is to be hoped that the zenith of the testing culture will soon be a thing of the past. (Stern, 1930, pp. 50–51)

Far from soon becoming a thing of the past, the testing culture that so concerned Stern in 1930 has become, thanks in no small part to the work of U.S. psychologists, far stronger than it was even then, and I can see no evidence that critiques of that culture, such as one mounted nearly one quarter of a century ago by the anthropologist F. Allan Hanson, have had any lasting effect at all. As Hanson (1993) argued:

America is awash in tests ... . The test giver—that institutional entity, whatever it may be, that we refer to under the name of 'they'—acquires almost godlike status, and tests become instruments of social regimentation and discipline. (From the dust jacket of Hanson's 1993 book)

In this light, I find even more haunting the words with which Stern concluded his 1931 presentation in Moscow:

I come to the end. As a conclusion, it is perhaps not idle to point out that the theme of my lecture has been ... the personal factor in psychotechnics. For me, the question has simply been: What does psychotechnics, and practical psychology, mean for the individual person who is subjected to the methods of those disciplines? It is because this question is usually pushed to the background in discussions of the nature and significance of psychotechnics that I thought I should give it special attention. (Stern, 1933, p. 63)

Conclusion

When in 1931 Stern spoke the words 'I come to the end' as he brought his Moscow presentation to conclusion, he could not have known how tragically prescient he was. Not fully two years later, in January of 1933, Hitler and the Nazis came to power in Germany, and Stern was soon banned from all of his teaching, research, and administrative activities at the University of Hamburg. In a letter dated April 27 of that year, Stern submitted his formal resignation, and his scholarly life was virtually ended.

He fled first to the Netherlands, where he did manage to complete work on his *General Psychology from the Personalistic Standpoint,* which was published in German by the Dutch press Martinus Nijhoff in 1935 (Stern, 1935). An English translation of the work by Howard David Spoerl was published in 1938 by Macmillan, New York (Stern, 1938). Meanwhile, Stern had proceeded to the U.S., where he accepted a faculty position at Duke University. Professionally speaking, the few years that Stern was at Duke prior to his death in 1938 were, by his own standards, not very productive ones.

In the introductory pages of the *General Psychology* text just mentioned, Stern reiterated his long-held conviction that the discipline of psychology could not sensibly divorce itself from philosophy. He wrote:

> The separation of two independent approaches—the metaphysical and the empirical—is no more possible within scientific psychology than in lay psychology or artistic psychology. On the contrary, a symbiotic relationship between philosophical considerations and methodological findings is unavoidably necessary. The conviction, still now widespread – that psychology could or should become a discipline fully independent of philosophy leads either to a psychology without a psyche or to scientific work that incorporates a world view and grounding epistemological presuppositions that are not consciously recognized. (Stern, 1938, p. 10)

Stern's central point in this passage is one that he expressed in a letter that he had written to Jonas Cohn more than three decades earlier, on November 11, 1900. In that letter, Stern noted that he had been 'reading Münsterberg'—presumably the *Grundzüge der Psychologie* (Foundations of Psychology), which Münsterberg had published earlier that year (Münsterberg, 1900). Stern indicated that although he was favorably impressed by that work in many respects—he wrote that 'Münsterberg was better than his reputation'—he was still left unsatisfied by Münsterberg's philosophical two-sidedness. 'One cannot,' Stern wrote to Cohn, 'be an ethical idealist in metaphysics and a mechanist in psychology.'

Stern was concerned throughout his scholarly life to overcome such two-sidedness. Critical personalism was, in his view, a genuine *Weltanschauung* or comprehensive worldview—such thinking seems decidedly anachronistic to us today—within the framework of which his critically *inter*-personal sensibilities and, indeed, the entire socio-cultural outlook he favored, were intimately bound up with his scientific conception of the nature of persons.

Critical personalism insists on irreducible distinction between *persons* and *things*. From the standpoint of critical personalism, therefore, a 'basic' scientific psychology that represents persons as if they were things errs epistemologically, and an 'applied' psychology that endorses the regard for and treatment of persons as if they were things is socio-ethically problematic. In various ways, the works by Stern that have been discussed in this chapter project the seamlessness with which he coordinated these two broad concerns, and it is arguable that contemporary psychology—both basic and applied—could benefit from greater familiarity with Stern's thinking.

Upon Stern's passing in 1938, Gordon Allport (1897–1967) concluded his appreciation of Stern's accomplishments by forecasting that although critical personalism had not gained much attention during Stern's lifetime, that system of thought would 'eventually have its day, and its day would be long and bright' (Allport, 1938, p. 773). The present work has been written in the hope that, both in the realm of basic science and in the applied, socio-cultural domain, Allport's confident prediction will yet prove true.

Notes

1 A copy of this document was made available to me by Werner Deutsch (1947–2010). In the discussion of Stern's views on 'tolerance' that follows, all quotations are from that unpublished document.

2 My English translation of the 1909 monograph was published by APA Books in 1999 (C. Stern and W. Stern, 1999).

3 The question of just how one is to think about the 'distinctiveness' of individuality if not in a way that conforms to the underlying psychological tests will be discussed in Chapter 6.

4 The expression 'Taylorism' refers to an approach to worker management championed around the turn of the 20th century by Frederick Winslow Taylor (1856–1915). Taylor emphasized the utility of deploying workers in accordance with their differential skill sets, as ascertained by means of standardized tests, so as to maximize worker productivity and, in turn, corporate profits.

References

Allport, G. W. (1938). William Stern: 1871–1938. *The American Journal of Psychology, 51*, 770–773.

Feger, B. (1991). William Sterns Bedeutung für die Hochbegabungsforschung – die Bedeutung der Hochbegabungsforschung für William Stern (William Stern's significance for research on the highly gifted, and vice versa). In W. Deutsch (Ed.), *Die verborgene Aktualität von William Stern*, S. (pp. 93–108). Frankfurt am Main: Verlag Peter Lang.

Graf-Nold, A. (1991). Stern versus Freud: Die Kontroverse um die Kinder-Psychoanalyse - Vorgeschichte und Folgen (Stern versus Freud: The controversy over child psychoanalysis - pre-history and consequences). In W. Deutsch (Ed.), *Die verborgene Aktualität von William Stern*, S. (pp. 49–91). Frankfurt am Main: Verlag Peter Lang.

Hanson, F. A. (1993). *Testing testing: Social consequences of the examined life*. Berkeley, CA: University of California Press.

Lamiell, J. T. (2003). *Beyond individual and group differences: Human individuality, scientific psychology, and William Stern's critical personalism*. Thousand Oaks, CA: Sage Publications.

Lück, H. E., & Löwisch, D.-J., (Eds.) (1994). *Der Briefwechsel zwischen William Stern und Jonas Cohn: Dokumente einer Freundschaft zwischen zwei Wissenschaftlern* (Correspondence between William Stern and Jonas Cohn: Documents of a friendship of two scientists). Frankfurt am Main: Verlag Peter Lang.

Münsterberg, H. (1900). *Grundzüge der Psychologie* (Foundations of Psychology). Leipzig: Barth.

Stern, C., & Stern, W. (1918). *Die Kindertagebücher: 1900–1918*. William Stern Archive, Jewish National and University Library at Givat Ram, Jerusalem. (A digitalized version of the unpublished diaries, created under the direction of Werner Deutsch (1947-2010) can be found at the Max Planck Institute for Psycholinguistics at the University of Nijmegan, The Netherlands.)

Stern, C., & Stern, W. (1999). *Recollection, testimony, and lying in early childhood* (J. T. Lamiell, Trans.). Washington, DC: American Psychological Association Books.

Stern, W. (1900a). *Über die ethische Bedeutung der Toleranz (On the Ethical Significance of Tolerance)*. Unpublished paper, Breslau, Germany.

Stern, W. (1900b). *Über Psychologie der individuellen Differenzen (Ideen zu einer "differentiellen Psychologie")* (On the psychology of individual differences: Toward a differential psychology). Leipzig: Barth.

Stern, W. (1901). S. Freud: Die Traumdeutung. (Rezension) [S. Freud: The interpretation of dreams. A review], *Zeitschrift für Psychologie und Physiologie der Sinnesorgane, 22*, 13–22.

Stern, W. (1911). *Die Differentielle Psychologie in ihren methodischen Grundlagen* (Methodological foundations of differential psychology). Leipzig: Barth.

Stern, W. (1914). Die Anwendung der Psychoanalyse auf Kindheit und Jugend. Ein Protest. Mit einem Anhang von Clara und William Stern: Kritik einer Freudschen Psychoanalyse. [The application of psychoanalysis to children and adolescents. A protest. With an appendix by Clara and William Stern: Critique of a Freudian psychoanalysis.] *Zeitschrift für angewandte Psychologie und psychologische Sammelforschung, 8*, 71–101.

Stern, W. (1923). *Person und Sache: System der philosophischen Weltanschauung. Zweiter Band: Die menschliche Persönlichkeit,* dritte unveränderte Auflage (Person and thing: System of a philosophical worldview. Volume Two: The human personality, third unrevised edition). Leipzig: Barth.

Stern, W. (1929). Persönlichkeitsforschung und Testmethode (Personality research and the method of testing). *Jahrbuch der Charakterologie, 6,* 63–72.

Stern, W. (1930). Eindrücke von der amerikanischen Psychologie: Bericht über eine Kongreßreise (Impressions of American psychology: Report on travel for a conference). *Zeitschrift für Pädagogische Psychologie, experimentelle Pädagogik und jugendkundliche Forschung, 31,* 43–51 und 65–72.

Stern, W. (1933). Der personale Faktor in Psychotechnik und praktischer Psychologie (The personal factor in psychotechnics and practical psychology). *Zeitschrift für angewandte Psychologie, 44,* 52–63.

Stern, W. (1935). *Allgemeine Psychologie auf personalistischer Grundlage (General psychology from the personalistic standpoint).* Den Haag: Nijhoff.

Stern, W. (1938). *General psychology from the personalistic standpoint* (H. D. Spoerl, Trans.). New York: Macmillan.

5 Some Critically Personalistic Observations on Current Discussions of Racism in American Society

It was mentioned in the preface to this book that since shortly after the highly publicized murder of Mr. George Floyd by a Minneapolis police officer in May of 2020, I have been participating with nine other persons, five of them Black; the other four White, in regularly-held meetings to discuss the problem of racism in American history and in contemporary American society. In our early meetings, group members openly shared their personal experiences with, and current perspectives on, this topic. Discussions in subsequent meetings have been based primarily on our engagement with a wide variety of relevant materials—books, book chapters, magazine and newspaper articles, and video documentaries—pertaining to the history of racism and to the current state of interracial relations in the U.S. Through these efforts, every member of our group has acknowledged a broadening of perspective on the role of race in American history and on race relations in contemporary American society.[1]

As I hope to make clear in what follows, the topics I have singled out for commentary pertain to matters of fundamental importance, both conceptual and moral/ethical, to our understanding of what we are doing—and not doing—when we engage with one another about race and racial issues in contemporary society.

The Impersonal Nature of Discourse about Personkinds

Some General Considerations

The great preponderance of contemporary discourse about race, both within our discussion group and in the published materials with which we have been engaged, has actually been discourse not about *persons* but about *personkinds*—mostly about the personkinds 'Black' and 'White,' of course, but at times about other personkinds as well (e.g., Jews, Hispanics). Widespread belief has it that discourse about personkinds must also be, if only by default, discourse about the individual persons

DOI: 10.4324/9781003375166-7

who instantiate those personkinds. This belief is false, however, and it is arguable that the widespread failure to recognize this among professionals and laypersons alike is a major hindrance to efforts toward the mitigation of racism in contemporary society.

Without question, discourse about personkinds can be valuable, and even necessary in certain contexts. It provides what social critic Wendell Berry (b. 1934) calls 'public knowledge,' and it can serve to further public discourse (cf. Berry, 2022). It is important to recognize, however, that such knowledge, and the discourse that generates and is furthered by it, is about *populations as such*, and that it is guided by essentially *statistical* considerations. The discourse always refers, implicitly if not explicitly, to established empirical facts—or to their would-be proxies in the form of subjective opinions or beliefs—about the relative prevalence of certain phenomena (behaviors, thoughts, feelings, attitudes, beliefs, wants, intentions, etc.) within designated collectives of people, with those collectives being defined by some categorical distinction(s) such as race, sex, age, class, religious confession, political leanings, etc.

Consider a statement such as 'He's just a *typical* senior White male.' As it stands, that statement makes no explicit reference to any actual statistic. Nevertheless, it asserts an essentially statistical idea, namely, a state of affairs, 'typicality,' regarding the personkind 'senior White males.' Similarly, when a participant in a discussion asks 'Isn't the young urban Black person Smith like *most* young urban Black persons?' in some or other regard, that participant is posing a question of an essentially statistical nature: it asks for knowledge of, or belief concerning, what is true about the majority ('most') of the personkind 'young urban Blacks' in some regard. A statement such as 'Well, I think (or 'studies show') that a Hispanic person is *more likely* than a Caucasian person to do or think thus-and-so is tied to considerations of *probability*, considerations that are quintessentially statistical in nature.

It bears re-emphasis here that in all of these examples, and countless others that could be formulated, the discourse expresses essentially statistical notions about populations *whether or not reference is made to any actual statistic(s)*. Once this basic point is grasped, the ubiquity of statistical considerations about personkinds in discourse about race and other social issues is hard to miss.

This point is so important because although statistical thinking about personkinds is inherently *im*personal, it is easily disguised in language that makes it *seem* personal.

Statistical thinking is inherently impersonal because it allows no room for the notion that every person is individually distinct.[2] Instead, such thinking logically demands that every particular instance of a given personkind be regarded as substitutable, both *by* and *for*, every other particular instance of that same personkind. This logical demand is the very antithesis of individual distinctiveness. Thus, and however unwittingly, it reduces persons to things, a reduction fundamentally contrary to critically personalistic thinking (Lamiell, 2003).

The problem here is only exacerbated by language fostering the illusion that knowledge of personkinds is, at one and the same time, knowledge of the individual persons who instantiate those personkinds. That illusion is created and sustained by the inappropriate use of the concept of *probability*. Specifically, knowledge of the proportional prevalence of phenomenon X within a designated personkind is falsely interpreted as the equivalent of knowledge of the 'probability' ('likelihood,' 'chances') that phenomenon X is (or was, or will be) manifested by any given individual instantiating that personkind.

Trenchant warnings about the illogic of such interpretations were being sounded already in the 19th century. Nevertheless, public discourse about social issues in general is saturated with just such interpretations, and, sadly, they are lent the mantle of authority by trained social scientists who should know better. Given this fact, and given the importance of avoiding the reduction of persons to things that is inevitably entailed by personkind thinking, some acquaintance with the perspectives on this matter that have emerged historically will be helpful as backdrop for a critical discussion of the stance adopted by sociologist Robin DiAngelo (b. 1956) in her best-selling book *White Fragility: Why It's So Hard for White People to Talk About Racism* (DiAngelo, 2018).

We turn first to the ideas of the 19th-century Belgian polymath, Adolphe Quetelet (1796–1874).

A Brief Sojourn into Some Relevant History

According to the intellectual historian Theodore M. Porter (b. 1953), Quetelet was of the view that the doings of individual persons were too unpredictable, whether because of sheer capriciousness or because of the complexity of determining factors, to be a fit subject for scientific inquiry (Porter, 1986). Quetelet did believe, however, that a degree of order in human conduct sufficient to yield to the scientific quest for knowledge of lawful regularities could be discerned by studying *aggregates* of individuals statistically. He called his proposed discipline *social physics,* and made the conceptual cornerstone of that field the entity *l'homme moyen,* or *the average man.* Porter elaborated the meaning Quetelet wished to convey with that concept as follows:

> In principle, wrote Quetelet, the courage or criminality of a real person could be established if that person were placed in a great

number of experimental situations, and a record kept of the number of courageous or criminal acts elicited. This would be interesting, but it was wholly unnecessary for social physics. Instead, the [social] physicist need only arrange that courageous and criminal acts be recorded throughout society, as the latter already were [anyway], and then the average man could be assigned a '*penchant for crime*,' equal to the number of criminal acts committed, divided by the population. In this way, a set of discrete acts by distinct individuals was transformed into a continuous magnitude, the *penchant,* which was an attribute of the average man. (Porter, 1986, p. 53, italics in original; brackets added)

Quetelet understood that knowledge of populations was just that: knowledge of populations. While that knowledge could be conveyed by pithy references to a singular entity, *l'homme moyen,* those references would be to an *abstract, hypothetical* entity, and the knowledge of that entity was neither to be mis-cast by the knowledge purveyors nor mistaken by the knowledge consumers as knowledge about any actual flesh-and-blood individual. Relevant to this point, Porter (1986) quoted Quetelet (in translation) as follows:

If one seeks to establish, in some way, the basis of a social physics, it is *l'homme moyen* whom one should consider, *without disturbing oneself with particular cases or anomalies, and without studying whether some given individual can undergo a greater or lesser development in one of his faculties.* (Quetelet, 1935, as quoted in translation from the French by Porter, 1986, pp. 52–53, italics added)

For all intents and purposes, Quetelet's social physics was a *demography.* Its concern was with populations *as such*, and not with any particular someone within those populations. In effect, this means that the concern of Quetelet's social physics was with the production of knowledge about, quite literally, *no one.*

However, among turn-of-the-20th-century psychologists, there arose great eagerness to make their own discipline practically useful in the world outside the laboratories of the original experimental psychology.[3] Those psychologists found great use for the statistical methods of investigation proper to social physics. They saw that the skillful application of those methods in the analysis of data (e.g., scores on psychological tests) recorded for large numbers of research subjects properly 'sampled' from various populations would position psychologists to give practical advice to officials in schools, hospitals, commercial enterprises, the military, etc. on matters of substantive relevance to their respective institutions. This possibility opened up a professional niche that many early 20th-century psychologists with 'applied' interests were only too eager to fill (Danziger, 1990).[4]

Of course, psychologists using the statistical methods of social physics could not eschew claims of knowledge about individual members of the populations they were studying without undermining their own discipline. That is: had the thinking of those psychologists followed Quetelet's lead in this respect, they would have effectively been acknowledging that they were no longer doing *psychology* at all, but were instead prosecuting their own brand of demography, just as Quetelet himself knowingly was. Instead, however, the practice gained widespread favor among psychologists of interpreting the statistical relationships they were discovering—relationships that, logically speaking, were every bit the conceptual equivalent of Quetelet's *penchants*—not as empirical facts pertaining only to some fictional *l'homme moyen,* but rather as factual knowledge, however incomplete, about the psychological tendencies of real flesh-and-blood individuals within (or represented by) the studied populations (cf. Danziger, 1990; Lamiell, 2019). It is the presumed legitimacy of such interpretations that has long been—and continues to be—fueled by narratives of the sort mentioned earlier, according to which a statistical feature of a collective, e.g., 'Seventy-five percent of population X does A' may validly be said to mean that 'the probability is .75 that *"this"* member of population X does (or did, or will do) A.' It is via false claims of this very sort that statistical knowledge about populations is routinely made to *seem* as if it is knowledge of the individuals within or 'represented' by those populations.

One of the aforementioned warnings about the illogic of such interpretive practices was issued by the British philosopher and logician John Venn (1834–1923). Making clever use of an analogy to coin-flipping, Venn articulated his point as follows:

> I am about to toss up [a fair coin], and I therefore half believe, to adopt the current language, that it will [turn up heads]. Now it seems to be overlooked that if we appeal to the event, ... our belief must inevitably be wrong ... . For the thing must either happen or not happen: i.e., in this case, the penny must either [turn up heads or not]; there is no third alternative. But whichever way it occurs, our half-belief, so far as such a state of mind admits of interpretation, must be wrong. If [the flip turns up] heads, I am wrong in not having expected it enough; for I only half believed in its occurrence. If it does not happen, I am equally wrong in having expected it too much; for I half believed in its occurrence, when in fact it did not occur at all. *The same difficulty will occur in every case in which we attempt to justify our state of partial belief in a single contingent event.* (Venn, 1888, p. 141; brackets and emphasis added)

The crucial epistemic point with which Venn concluded this passage underscores emphatically the truth of an observation that had been made two decades earlier by the German philosopher and mathematician Moritz Wilhelm Drobisch (1802–1896). Drobisch's statement was quoted in Chapter 1, and I reintroduce the quote for emphasis here, noting that one has only to substitute the word 'aggregate' where the word 'average' appears—an 'average' being, technically, a particular kind of aggregate statistical index—in order to capture the wider applicability of the point Drobisch was making:

> It is only through a great failure of understanding that the mathematical fiction of an average [aggregate] man can be elaborated as if all individuals [within the aggregate] possess a real part of whatever obtains for this average [aggregate] person. (Drobisch, 1867, as quoted in Porter, 1986, p. 171, brackets added)

Applying Drobisch's point to the example immediately above, knowledge that 75% of population A does X does *not* mean, and may not validly be said to mean, that probability that any given member of population A does (or did, or will do) X is .75. Any given member of population A *either does* (did, will do) X *or does not* (did not, will not) do X. The statistical fact '.75' is a fact about a *series* of observations considered as a series, so, '75-out-of-100.' It is not a fact about *any* one observation within that series. It is in just this sense that a population-level statistical fact is *factual about no one.* The failure to grasp this point *is* the 'great failure of understanding' identified by Drobisch.

Unfortunately, the indulgence by psychologists (and many other social scientists) of just that 'great failure of understanding' has by now widely infected lay persons' narratives about people, too. The result of this development is an intellectual malady I have labeled '*statism*' (refer to Chapter 1), using this word to label that seemingly incorrigible faith in the power of aggregate statistical knowledge to advance our scientific understanding not only of population-level phenomena, as Quetelet had foreseen, but of individual-level doings as well (cf. Lamiell, 2019).

The spirit of statism, so completely contrary to the warnings issued by Drobisch and by Venn in the 19th century, and periodically by other thinkers since then (cf. Lamiell, 2019; Porter, 1986) is verily crystallized in the words of the British historian Henry Thomas Buckle (1821–1862), who wrote in *A History of Civilization in England,* first published in 1857:

> [From carefully compiled statistical facts] more may be learned [about] the moral nature of Man than can be gathered from all the accumulated experiences of the preceding ages. (Buckle, 1857/ 1898, p. 17)

Buckle was fully convinced that, in fact, aggregate-level statistical regularities *are* instructive not only about the doings of personkinds as collectives, but also—and, indeed, *therefore*—about the doings of individuals instantiating those personkinds (cf. Porter, 1986). It is that belief that bears the infection of statism, and it is that infection that so permeates contemporary discourse about personal and inter-personal doings generally. Nowhere in the materials pertaining to racism with which our discussion group has been engaged during the past three-plus years (refer above) has that infection been more prominent, or fomented more contentiousness within the group, than in social scientist Robin DiAngelo's (b. 1956) best-selling book *White Fragility: Why It's So Hard for White People to Talk About Racism* (DiAngelo, 2018).

Statism in White Fragility

Near the beginning of her book, DiAngelo, to her credit, spoke directly to the issue of our immediate concern. She did this by posing to herself, in wholly rhetorical fashion, the question of how she could claim to know something about the views on racism held by individuals whom she has never met. She answered her own question:

> As a sociologist, I am quite comfortable generalizing; social life is patterned and predictable in measurable ways ... . There are, of course, exceptions, but patterns are recognized as such precisely because they are recurring and predictable. (DiAngelo, 2018, pp. 11–12)

As I hope has been made clear by the foregoing discussion, the problem with DiAngelo's thinking on this matter lies in her failure to realize that the 'patterning and predictability' in social life to which she alludes is a *strictly population-level* phenomenon, knowledge of which entitles *no knowledge claims whatsoever* about *any* individual. As we have seen, Quetelet, Venn, and Drobisch were fully attuned to this point, while Buckle was not. Alas, by now, mainstream thinking among social scientists has aligned itself squarely with the views of Buckle. To repeat for emphasis a statement made earlier: *population-level knowledge is knowledge of no one.*

In the course of our discussion group's engagement with *White Fragility,* some members adopted the position that since they had encountered, first-hand, many individuals whose behavior did, in fact, match the generalization about White persons running through DiAngelo's entire narrative, her generalizations were justified. In fact, however, such encounters do not justify DiAngelo's *a priori* generalizations, however much they might seem to do so on the surface.

Admittedly, the point here is a subtle one, which perhaps explains why it has proven so difficult for contemporary thinkers to grasp. Nevertheless, the matter is of sufficient importance that the effort to achieve clarity on it must be furthered.

First off, it is important to see that the crucial question begged by DiAngelo's generalizations in *White Fragility* is not: Can White persons be identified who do, in fact, have difficulty talking about racism? To be sure: if the experience of such a difficulty is, as DiAngelo contends, widely prevalent among White people—and I know of no basis for challenging DiAngelo's claims on this matter—then repeatedly pre-suming ahead of time that any given White individual will experience difficulty discussing racism would ensure that, *in the long run*, one's presumptions will be confirmed more often than contradicted. But the proviso 'in the long run' is a strictly *population*-level criterion for deciding the serviceability of a strategy for pre-judging individual cases so that, ultimately, one's judgmental errors will be minimized. The individual-level mistakes that will inevitably occur are reduced to acceptable collateral damage. Thus, the actual concern here is not with what is factually the case about each individual being judged, but rather with what is best *for the decision-maker* to presume *a priori* about individual cases in order to minimize his or her errors of judgment in the long run.

From a personalistic standpoint, the question that is actually begged by DiAngelo's (2018) generalizations in *White Fragility* is: based on what is taken to be statistically true of White people as a collective when it comes to talking about (dis)comfort in discussions of racism, what may one validly presume to be the case about the (dis)comfort level that *this* White person under immediate consideration will experience, or has experienced, upon entering into a discussion of racism? The only correct answer to this question is 'Nothing,' and that is the only correct answer for *every* individual White person about whom the question might be posed. This is *because* population-level knowledge is knowledge of no one.

The disclaimer in DiAngelo's statement that 'there are, of course, exceptions' to the rule defined by the statistical regularity on which she bases her generalizations seems at first blush to be a forthright expression of cautiousness, warranted by her certainty that lurking somewhere within the membership of the personkind 'White people' are individuals who, contrary to the book's stereotypical conceit, will prove to be quite comfortable talking about racism. Logically, however, such caveats are vacuous. This is because the very sensibility of regarding some particular case as an 'exception to a rule' entails the assumption that the rule in question pertains to individual-level doings in the first place. As we have seen, this assumption does not hold for aggregate-level statistical 'rules' defined only for populations, and, logically, individual cases cannot be regarded as 'exceptions' to—or, for that matter, affirmations of—'rules' that, precisely because of their aggregate, statistical nature, are not rules about individual-level doings to begin with. In effect, therefore, an *a priori* disclaimer that one should expect individual exceptions to empirical regularities that are not even defined at the level of the individual amounts to a disingenuous attempt to justify the continued exercise of such rules to pre-judge individual cases while promising to later rescind the inevitably false pre-judgments of some of those cases with a perfunctory 'oops.'

The inferential constraint logically demanded by the certainty of individual exceptions to population-level statistical rules is a total one. Since, in practice, *every* individual case, considered as such, *could* prove to be one of the 'exceptions to the rule,' the non-prejudicial consideration of each individual case requires that the statistical 'rule' *always* be ignored![5]

Once again, we find ourselves face-to-face with the reality, as counter-intuitive as it is to many, that knowledge about personkinds is *not* knowledge of persons. The epistemic imperative here calls for more than an acknowledgment that the knowledge contained in a statistical regularity defined for a population may not be presumed to hold for *every*one in that population. The imperative is to recognize that such knowledge may not be *presumed* to hold for *any*one in that population. In failing to grasp and scrupulously respect this epistemic imperative, safe harbor is offered for exercises in stereotyping. Let there be no mistake: this is no less true when generalizations are guided by statistical regularities established by methodologically sophisticated social science than when they are guided by flimsy and baldly irresponsible prejudices toward particular personkinds. Stereotyping is stereotyping, and we do well to bear in mind that exercises in stereotyping are a core component of racist thinking. Beyond the epistemic imperative, there is a moral/ethical imperative here as well.

In one of our discussion group meetings devoted to *White Fragility,* one member of the group took the occasion to read a passage from the book, and then, taking umbrage at having. been collaterally damaged by DiAngelo's exercises in stereotyping, expressed thorough disdain for the book by throwing it to the floor. Not all members of the group reacted approvingly to this display. But the very fact of our group's existence reflected the eagerness of members, Black and White alike, to enter into discussion of racism, and thus belied the validity of DiAngelo's generalizations about the White members of the group. Moreover, the emphatic display of displeasure with the book by one member of the group was in part an expression of the realization that DiAngelo's stereotyping would falsely pre-judge many individuals beyond the confines of our group.

DiAngelo herself acknowledged that *White Fragility* had alienated a substantial number of her White readers, an effect that would seem counter-productive to her larger goal of recruiting White persons to the much-needed effort to mitigate the effects of racism—especially of the systemic sort—in contemporary society. In this light, perhaps, the potential merits of a more personalistic approach to discussions of racism, and the hazards of discourse guided entirely by uncritical, impersonalistic, personkind thinking, can become more apparent.

Other Impersonal Aspects of Contemporary Discourse about Race in America

Arguably, a clear understanding of the intellectual malady of statism, as discussed above, is the single most pervasive way in which contemporary discussions of social issues in American society could be advanced by the adoption of a critically personalistic perspective. This is why I have devoted as much of this chapter as I have to the matter. There are other issues, however, and in what follows I will mention three domains in which current discourse could benefit from critically personalistic thinking.

Mechanistic Narratives Concerning the Social Dynamics of Systemic Racism

One of these domains is encountered in narratives that portray persons as entities whose psychological doings—again: perceptions, judgments, emotions, cognitions, attitudes, behaviors, etc.—are extensions of the complex entirety of their respective socio-cultural histories.[6] A vivid example of what I am referring to here is found in the excellent book by the Pulitzer Prize-winning journalist Isabel Wilkerson (b. 1961) titled *Caste: The Origins of Our Discontents* (Wilkerson, 2020).

In her book, Wilkerson quite properly deplores the widespread failure within contemporary American society to appreciate the existence and lasting untoward effects of *systemic* racism, that form of racism found not in the blatant actions of isolated individuals, but in the fundamentally racist structure of many of our civic institutions and practices. Continuing to understand racism as a strictly individual-level phenomenon, Wilkerson argues, 'keeps the focus on the single individual, rather than *on the system that created that individual*' (Wilkerson, 2020, p. 69).

First of all, it would be quite erroneous to contend that a fully coherent account of the causal dynamics of systemic racism could somehow bypass the consideration of individual persons. It may very well be that, to invoke an increasingly popular metaphor, the systemic form of racism is 'baked in' to the 21st-century American culture.

However, this does not mean—*could* not mean—that systemic racism has no presence in individual psyches, for individual psyches are the very 'pans' in which culture is metaphorically 'baked.' Theoretical recognition of this proviso is necessary for any coherent account of the causal effects of systemic racism on community life. The reason for this was once succinctly expressed by the late Oxford philosopher and social psychologist Rom Harré (1927–2019): 'Causal processes occur only in individual beings, since mechanisms of action, *even when we act as members of collectives,* must be realized in particular persons' (Harré, 1981, p. 14, emphasis added).

A critically personalistic perspective on systemic racism will not fail to respect Harré's insight here. In doing so, however, it will not embrace an understanding of persons simply as the 'products' or 'creations' of the society into which they happen to have been born and within which their lives happen to be unfolding. Such a view preempts consideration of the agency of persons themselves in the causal complex that undeniably includes but is not exhausted by the effects, be they conscious or unconscious, of institutional beliefs and practices on individual doings—deliberately racist or otherwise.

The agency postulated by critically personalistic thinking is understood to be exercised by the capacity persons have, as *valuational* beings, to appraise and to then *adopt or reject* the values pressed upon them by prevailing socio-cultural teachings and practices. It would appear that this view was shared by the renowned writer and civil rights activist James Baldwin (1924–1987). According to Eddie S. Glaude, Jr., Professor of African American Studies at Princeton University and Baldwin scholar, Baldwin

> insisted that we are not the mere product of social forces. Each of us has a say in who we take ourselves to be. No matter what America said about him as a black person, Baldwin argued, he had the last word about who he was as a human being and as a black man. (Glaude, 2020, p. 37)

Absent the capacity to 'have a say in who one is as a human being,' there is no firm conceptual basis for attributing to anyone any measure at all of moral responsibility for anything: neither for the existence, persistence, and modification of one's own racist beliefs and practices as an individual, nor, as a citizen, for the perpetuation—even if not the establishment—of racist structures and practices now institutionalized in the communities of which she or he is a member.

Elsewhere in *Caste,* Wilkerson clearly recognizes the validity of these moral imperatives. The following passage beautifully expresses her convictions in this regard:

None of us chose the circumstances of our birth. We had nothing to do with having been born into privilege or under stigma. We have everything to do with ... how we treat others in our species from this day forward. We are not personally responsible for what people who look like us did centuries ago. But we are responsible for what good or ill we do to people alive with us today. We are, each of us, responsible for every decision we make that hurts or harms another human being. We are responsible for recognizing that what happened in previous generations at the hands of or to people who look like us set the stage for the world we now live in ... (Wilkerson, 2020, pp. 387–388)

The point of the foregoing is to underscore the importance of avoiding discourse that portrays persons as entities whose doings are determined (i.e., 'produced' or 'created') by their social circumstances. Such discourse effectively reduces persons to things, and things are entities not subject to any moral imperatives at all. Such considerations lie in the domain of 'should,' and as Stern advised in a passage cited in Chapter 2, mere things cannot 'should.'

Impersonal Understandings of the Concept of Personal Identity

In the domain of social discourse, the notion of 'personal identity' is widely understood to refer to that which is distinctively characteristic of a person. In practice, the concept of personal identity is commonly folded into discourse about social interactions by characterizing individuals in accordance with the personkind categories relevant to the substantive concerns of the discourse. If, for example, the discourse is about racial matters, then it is each individual's race that is viewed as distinctively characteristic about him or her for the purpose(s) of that discourse.

Inevitably, this framework for 'capturing' personal identities involves stereotyping of the very sort discussed earlier in this chapter, and, like all stereotyping, it inevitably obscures possibly distinctive and consequential features of an individual who has been lumped together with others 'of like kind' for the purposes of the discourse.

It was out of concern for just such obfuscation that Kimberlé W. Krenshaw (b. 1959) introduced the concept of 'intersectionality' into discussions about race and racial issues. Employing this concept, Krenshaw has hoped to refine understandings of various issues relevant to race in America by considering simultaneously multiple criteria for capturing individuals' identities (read: for more nuanced categorization of them). In one of her works, for example, Krenshaw (2016) has sought to explain how the consideration of race, in addition to the consideration of gender, can refine our understanding of the social problem of violence perpetrated against women. Her analysis suggests that a finer understanding of that problem can be achieved if the race of victimized women is taken into account rather than ignored in deference to the consideration of gender alone.

It is not difficult to see that the very problem that results whenever knowledge of persons is sought through the study of personkinds, i.e., the problem of obscuring aspects of the individuality of those who are being regarded as identical for purposes of the study, remains no matter how many intersecting personkinds are taken into consideration. There will always be finer distinctions to make when persons are categorized in terms of personkinds, and the only way to circumvent that problem altogether is to abandon completely the practice of regarding persons as instantiations of personkinds. The alternative is to embrace a view of persons as the individually distinct entities that they are seen to be from a critically personalistic perspective.

How to understand personal identities in this way? The key to answering this question lies in fully appreciating the critically personalistic tenet that human persons are irreducibly *valuational* in nature. From a critically personalistic perspective, the course of every individual's life is guided ultimately by the value that that person places on certain ends or goals. This leads to the realization that, when all is said and done, what is distinctively characteristic about a given individual, and therefore irreducibly definitive of that individual's personal identity, is *the set of values according to which that individual is conducting his or her life.*[7] Put simply, on this view one's identity as a person is defined by the values that guide one's doings.

In sharing this idea with the members of my racial issues discussion group, I have encountered pushback in the form of the question: Can we not, in the spirit of the just-noted observation by Krenshaw (refer to endnote 7), infer the values that are guiding an individual's life based on what we know are 'regularly' or 'typically' the guiding values among persons of that individual's kind? For reasons I hope to have made clear through the discussion earlier in this chapter, the critically personalistic answer to this question is and must be 'no.' Knowledge of the regularity with which certain values are held by members of some personkind (which could be defined as the 'intersection' of several personkinds) is population-level knowledge of an essentially statistical nature, and we have already considered both that and why such knowledge is *never* a sufficient basis for pre-supposing what values are held by *any* individual instantiation of that personkind. To lose sight of this logical truth and allow inferences of the sort recommended in the rhetorical question just posed is to court yet again Drobisch's (1867) 'great failure of understanding' and thus fall back on stereotyping. Conversely, to bear that logical truth in mind is to see (again) why the pursuit of knowledge about persons requires the complete abandonment of personkind—i.e., stereotypical—thinking, however daunting that prospect may be to contemporary sensibilities.

In the work cited just above, Krenshaw (2016) draws a distinction between social identity and personal identity that should not be overlooked here:

> We all can recognize the distinction between the claim 'I am Black' and the claim 'I am a person who happens to be Black.' 'I am Black' takes the socially imposed identity and empowers it as an anchor of subjectivity. 'I am Black' becomes not simply a statement of resistance but also a positive discourse of self-identification, intimately linked to celebratory statements like the Black nationalist 'Black is beautiful.' 'I am a person who happens to be Black,' on the other hand, achieves self-identification by straining for a certain universality (in effect, 'I am first a person') and for a concomitant dismissal of the imposed category ('Black') as contingent, circumstantial non-determinant. There is truth in both characterizations, of course, but they function quite differently depending on the political context. At this point in history, a strong case can be made that the most critical resistance strategy for disempowered groups is to occupy and defend a politics of social location rather than to vacate and destroy it. (Krenshaw, 2016, pp. 248–249, parenthesesin original)

The socio-political function of personkind thinking which Krenshaw (2016) exemplifies here with the expression 'I am Black' is an important one, and in the context of discourse about the current state of racial issues in American society, such thinking is arguably indispensable. This point will be further elaborated in the following discussion of the metaphor of 'colorblindness' in societal doings. This does not, however, gainsay the argument that when concerns shift from socio-political identity to personal identity, abandoning personkind thinking in favor of a critically personalistic focus on individuals' life-guiding values is necessary. This crucial point is vividly underscored in the final chapter of the book by the philosopher Susan Neiman (b. 1955) titled *Learning from the Germans: Race and the Memory of Evil* (Neiman, 2019). Questioning the long-term viability of identity politics in mitigating inter-racial tensions in American society, Neiman, an American-born Jew who for much of her life has chosen to reside in Berlin, related a lesson drawn from her own life. Using the expression 'tribalism' to refer to what I have called 'personkind thinking' about personal identities, Neiman wrote:

For five years of my own life, I gave tribalism a try. Soon after the 1995 Oslo Accords, I moved [from Berlin] to Israel ... In becoming an Israeli citizen, I tried out voting for tribalism with my feet. There I learned I could not possibly feel more connected to an arms dealer who shared my ethnic background than to a friend from Chile or South Africa or Kazakhstan *who shares my basic values*. My ties are to agents, not genealogies.

I choose friends, and loves, for reasons. (Neiman, 2019, pp. 383–384, brackets and emphasis added)

Neiman (2019) concluded her book by characterizing it

... as an exercise in universalism ... . If we fail to understand that we have more in common than all that divides us, we cannot pursue what Toni Morrison [1931–2019] called the human project, [which is] 'to remain human and to block the dehumanization and estrangement of others.' (Neiman, 2019, p. 384, brackets added)

I think it is not insignificant that Neiman's appeal to universalism converges with Krenshaw's (2016) allusion to the same theme in her elaboration of the expression 'I am a person who happens to be Black' (refer above). Fundamental to universalism, and hence prerequisite for the pursuit of Toni Morrison's 'human project' with any hope at all of success are (1) an unwavering commitment to the distinction between persons and things, so as to block the dehumanization of persons, and (2) a sound conceptual alternative to personkind thinking—which itself blurs the person-thing distinction—in order to block the estrangement from persons that results when they are viewed through the ever foggy lenses of impersonal stereotypes.

These considerations lead us directly to the last topic to be mentioned in this chapter, namely, the topic of 'colorblindness.'

On the Possibility of Colorblindness in Contemporary Social Exchanges

In perhaps the most widely-known line in the Rev. Dr. Martin Luther King Jr.'s (1929–1968) 'I have a dream' speech, delivered in Washington, D. C. on August 28, 1963, he expressed his hope that, one day, his four little children would live in a nation where they would be judged not by the color of their skin, but instead by the content of their character. In time, that passage would become an iconic paean to the ideal of a *'colorblind'* society, i.e., one finally worthy of the universalistic aspirations of 'liberty and justice for all.'

Without question, a society in which every person would be treated according to the content of his or her character, i.e., the core values guiding that person's conduct, would be consistent with the aspirations of critically personalistic thinking. Nor is this a mere coincidence. For even if the Rev. Dr. King had no direct familiarity with William Stern's critical personalism, and I have no reason to believe that he did, he benefited in his doctoral studies at Boston University from the tuition of prominent philosopher/theologians Edgar A. Brightman (1884–1953) and L. Harold DeWolf (1905–1986), both of whom embraced the tenets of personalistic thinking.[8]

In contrast to King's views, a society that would condone the differential allocation of basic rights to individuals based solely on personkind considerations of one sort or another—e.g., skin color, sex, socio-economic status, age, religious confession, etc.—would be antithetical to personalistic thinking: any *a priori* segmentation of a society into a hierarchy of personkinds would subvert the distinction between the inherent value of all persons and merely contingent value of things.

Clearly, the arc of personalistic thinking bends inexorably in the direction of colorblindness, and this is a philosophical stance that King both acknowledged and embraced. However, King also recognized—and famously reminded us—that though the more encompassing 'arc of the moral universe' bends ever toward justice, it is a long arc, indeed.[9] So while he was able to embrace the notion of colorblindness as socio-ethical ideal, he could not embrace it as a practical tool for achieving legal and political racial equality in his time (cf. Sundstrom, 2018).

To many, myself included, the present seems even less suited than King's time to colorblindness. The concern is that professions of colorblindness in one's social doings, even if sincere, will serve to rationalize looking past contemporary manifestations of *systemic* racism. This would undermine efforts now calling for the color *consciousness* necessary in order to redress the enduring legacies of past injustices perpetrated at a time when societal doings were anything but colorblind (cf. McGhee, 2021, esp. pp. 228–231).

It has also been argued that color consciousness rather than colorblindness among Blacks and other persons of color is a necessary instrument of racial celebration in the collective struggle to shatter the myth of White supremacy that has permeated American culture for many decades. This view aligns with Krenshaw's (2016) position regarding political identity, discussed above. It is also the view that was held by James Baldwin in advocating not for immediate colorblindness nor for unending color consciousness, but for color *transcendence*. As Glaude (2020) has characterized Baldwin's convictions in this regard, 'one can only transcend color by passing through it, and uprooting the lie [of white supremacy] along the way' (Glaude, 2020, p. 26). To transcend color differences in contemporary American society, then, would not be to ignore them, but rather to accord them the consideration that history demands they now receive, toward the ultimate goal of redressing the inequitable legacies of that history. Only in this way can a society that is color-transcendent and thus equitable going forward be possible.

Without doubt, my participation in the bi-racial discussion group to which I have referred in several places throughout this work has helped me to see that color transcendence, as understood by Baldwin in the fashion just described, is the critically personalistic aspiration appropriate for our present historical circumstances.

Notes

1 It pleases me greatly to be able to say that, as word about our discussion group began to circulate within our residential community, many additional residents beyond our 'core 10' expressed interest in being part of such a group, and, as of this writing, three additional groups of about the same membership size, with the same basic mission, have formed.

2 Instead of the expression 'individually distinct,' one might expect to find the word 'unique' in this context. I am deliberately avoiding the use of that term here, for reasons that will be elaborated in Chapter 6.

3 It will be recalled from Chapter 1 that in those laboratories, population-level statistical methods of inquiry had no place.

4 It happens that the overall structure of scientific psychology's new sub-discipline, called 'differential' psychology, was provided by none other than the architect of critical personalism, William Stern (cf. Stern, 1900, 1911). In time, Stern would become a harsh critic of the direction in which 'differential' psychologists were taking the discipline (Stern, 1929, 1930, 1933; see Lamiell, 2019, esp. Chapter 3 for a further discussion of this point).

5 Nearly 30 years ago, I applied this same thinking in my critical review of the book *The Bell Curve* (Herrnstein & Murray, 1994). See Lamiell (1996).

6 In more general terms, the causes of persons' doings are seen to lie in some combinations of their biological 'nature' and their socio-cultural 'nurture.' Little credence seems to be placed any more in the notion that racial beliefs and prejudices have anything to do with persons' biological natures, and so in this domain, the causal forces are now seen to lie in their socio-cultural 'nurture,' or what is commonly referred to as 'upbringing.'

7 Significantly, Krenshaw (2016) notes that the crux of matters in discussions of social issues that invoke the concept of intersectionality is not really to be found in the person categories themselves, on which the use of the concept relies, 'but rather [in] the particular values attached to [those categories]' (Krenshaw, 2016, p. 248, brackets added). So, in Krenshaw's view, too, it would seem, that values are fundamental to personal identities, and knowledge of an individual's personkind(s) is serving as a (less than satisfactory) proxy for knowledge of that individual's values.

8 Soon after earning his doctorate in 1955, King wrote to DeWolf: 'Both your stimulating lectures and your profound ideas will remain with me so long as the cords of memory shall lengthen. I have discovered that both theologically and philosophically much of my thinking is DeWolfian' (letter from M. L. King, Jr., to L. H. DeWolf, 2 June 1955; https://kinginstitute.stanford.edu/encyclopedia/dewolf-l-lotan-harold).

9 King expressed this view in a speech titled 'Remaining Awake Through a Great Revolution,' given at the National Cathedral in Washington, D. C., on March 31, 1968, just days before he was assassinated in Memphis, TN.

References

Berry, W. (2022). *The need to be whole: Patriotism and the history of prejudice.* Shoemaker and Company Publishers.

Buckle, H. T. (1857/1898). *A history of civilization in England.* New York: D. Appleton and Company.

Danziger, K. (1990). *Constructing the subject: Historical origins of psychological research.* New York: Cambridge University Press.

DiAngelo, R. (2018). *White fragility: Why it's so hard for white people to talk about racism.* Boston: Beacon Press.

Drobisch, M. W. (1867). *Die moralische Statistik und die menschliche Willensfreiheit: Eine Untersuchung* [Moral statistics and human free will: An investigation]. Leipzig.

Glaude, E. S., Jr. (2020). *Begin again: James Baldwin's America and its urgent lessons for our own.* New York: Crown.

Harré, R. (1981). The positivist-empiricist approach and its alternative. In P. Reason & R. Rowan (Eds.), *Human inquiry: A sourcebook of new paradigm research* (pp. 3–17). New York: Wiley.

Herrnstein, R. J., & Murray, C. (1994). *The bell curve: Intelligence and class structure in American life.* New York: The Free Press.

Krenshaw, K. W. (2016). Mapping the margins: Intersectionality, identity politics, and violence against women of color. In E. Taylor, D. Gillborn, & G. Ladson-Billings (Eds.), *Foundations of critical race theory in education,* second edition (pp. 223–250). London, UK: Routledge.

Lamiell, J. T. (1996). Hitting the curve. *Theory and Psychology, 6,* 317–322.

Lamiell, J. T. (2003). *Beyond individual and group differences: Human individuality, scientific psychology, and William Stern's critical personalism.* Thousand Oaks, CA: Sage Publications.

Lamiell, J. T. (2019). *Psychology's misuse of statistics and persistent dismissal of its critics.* London, UK: Palgrave-Macmillan.

McGhee, H. (2021). *The sum of us: What racism costs everyone and how we can prosper together.* New York: One World.

Neiman, S. (2019). *Learning from the Germans: Race and the memory of evil.* New York: Farrar, Strauss, and Giroux.

Porter, T. M. (1986). *The rise of statistical thinking: 1820–1900.* Princeton, NJ: Princeton University Press.

Stern, W. (1900). *Über Psychologie der individuellen Differenzen (\Ideen zu einer "differentiellen Psychologie")* [On the psychology of individual differences (Toward a "differential psychology")]. Leipzig: Barth.

Stern, W. (1911). *Die Differentielle Psychologie in ihrer methodischen Grundlagen* [Methodological foundations of differential psychology]. Leipzig: Barth.

Stern, W. (1929). Persönlichkeitsforschung und Testmethode [Personality research and the methods of testing]. *Jahrbuch der Charakterologie, 6,* 63–72.

Stern, W. (1930). Eindrücke von der amerikanischen Psychologie. Bericht über eine Kongreßreise [Impressions of American psychology: Report after travel to a conference]. *Zeitschrift für Pädagogische Psychologie, experimentelle Pädagogik und Jugendkundliche Forschung, 31,* 43–51 und 65–72.

Stern, W. (1933). Der personale Faktor in Psychotechnik und praktischer Psychologie [The personal factor in psychotechnics and practical psychology]. *Zeitschrift für angewandte Psychologie, 44,* 52–63.

Sundstrom, R. R. (2018). The prophetic tension between race consciousness and the ideal of color-blindness. In T. Shelby & B. M. Terry (Eds.), *To shape a new world: Essays on the political philosophy of Martin Luther King, Jr.* (pp. 127–145). Cambridge, MA: Harvard University Press.

Venn, J. (1888). *The logic of chance.* London/New York: Macmillan.

Wilkerson, I. (2020). *Caste: The origins of our discontents.* New York: Random House.

6 Toward a Broadened Perspective

Navigating Some Conceptual Obstacles to Critically Personalistic Thinking

In the previous chapter, I sought to provide readers with a preliminary sense for the applicability of critically personalistic thinking to some prominent features of contemporary discourse on race and racism in American society. In this final chapter, I point to certain ways in which the critically personalistic perspective could inform the manner in which we relate to—i.e., think about, speak to, and otherwise engage with—one another more generally. The discussion incorporates, for purposes of contrast, a consideration of how, at present, *non*-personalistic views are being reflected and perpetuated, wittingly or otherwise, in many of our social exchanges.

Full Disclosure: Confession of a Concrete Idealist

In the third and last volume of Stern's *magnum opus, Person and Thing*, i.e., the volume published in 1924 and titled (in translation), *'Philosophy of Value* (Stern, 1924), Stern characterized critical personalism as a 'concrete idealism,' distinguishing it from the 'abstract idealism' of the renowned German philosopher Georg Wilhelm Friedrich Hegel (1770–1831). Stern was advancing the idea that historically valued human ideals such as *truth* and *justice* are not merely unreachable abstractions but proximate and objective societal goals concretely realizable through the deliberate doings of persons in concert with one another.[1]

Whether a philosophical idealism, however 'concrete' it may be, has any genuine prospects for widespread adoption in modern times is today very much an open question (see, e.g., Neiman, 2023). However, whether critical personalism is *worthy* of thoughtful consideration as a framework for conducting and understanding social life is a matter on which I, in concert with the convictions of William Stern himself, have no doubt. Indeed, I believe that the need for such a framework, and for the spirit of universalism which it accommodates, is now as urgent as ever.

DOI: 10.4324/9781003375166-8

Some Prominent Obstacles to Critical Personalism—Then and Now

Early 20th-Century Behaviorism

Early in the 20th century, when Stern's scholarly works were being published (primarily by the Johann Ambrosius Barth press located in Leipzig, Germany), a major obstacle to the establishment of critically personalistic thinking within scientific psychology was that school of thought known as *behaviorism*. Following the lead of Ivan Pavlov's (1849–1936) famous laboratory experiments on the classical conditioning of salivation responses in dogs, the American psychologist John B. Watson (1878–1958) championed behaviorism as a paradigm for understanding the doings of all organisms—including but not limited to human beings—as reflex-like 'responses' fully determined by the specific conditions of stimulation prevailing upon the organism at any given moment. It was the thoroughly mechanistic nature of the behavioristic view of human doings—a view that effectively reduced persons to automaton-like things, always and ever just 'responding' to 'stimuli'—that made that view fundamentally irreconcilable with critical personalism (cf. Watson, 1913).

The widespread appeal of behaviorism within American psychology during Watson's time, and then through much of the career of B. F. Skinner (1904–1990), was a major factor in the obscurity of critical personalism throughout that epoch. But there is more to the story of that continuing obscurity, because behaviorism's rise was both coincident and conceptually compatible with a broader movement within psychology to wean the field from its mother discipline, philosophy. Watson himself embraced this development:

> With the behavioristic point of view now becoming dominant, it is hard to find a place for what has been called philosophy. Philosophy is passing—has all but passed, and unless new issues arise which will give a foundation for a new philosophy, the world has seen its last great philosopher. (Watson, 1928, p. 14)

Contra Watson's brash pronouncement, 'what has been called' (!) philosophy had not then, *circa* 1930, and has not yet, nearly a full century later, 'passed,' though behaviorism's domination of mainstream thinking in psychology certainly has. Unfortunately, the untoward effects of psychology's split from philosophy would endure, and they remain to this day—effects that were fully anticipated by two of the relatively few early critics of the movement favoring that split, namely, Stern and Wilhelm Wundt (1832–1920), experimental psychology's acknowledged founding father. Those two shared the view that philosophical thought is essential for effectively addressing questions of fundamental importance to scientific psychology that are *conceptual* in nature, and hence beyond the reach of any form of empirical investigation, including controlled experimentation and demography. Presciently, Stern maintained that psychology disengaged from conceptual questions would gradually deteriorate into a discipline of inferior intellectual quality (cf. Stern, 1917/2010; 1938), and Wundt argued that scientific psychology split off entirely from philosophy would eventually cease to exist (Wundt, 1913/2013).

Fatefully, and heedless of Stern's and Wundt's warnings, scientific psychology in the early 20th century did proceed during those formative years to distance itself from philosophy. Over time, the discipline has become ever more indifferent to conceptual/philosophical questions (Machado & Silva, 2007) and, sadly, ever less capable of handling them (Gantt & Williams, 2018). Of particular concern to us here is the continuing paradigmatic blindness within the field to the conceptual confusions lying at the heart of what I have throughout this work referred to as *statism*. Hindered by that blindness, legions of statistically oriented social scientists posturing as psychologists but functioning as demographers have managed to unwittingly transform what once was genuine psychology into what is now the discipline that I have termed 'psycho-demography' (see Lamiell, 2018, 2019).

Contemporary Statism

Arguably, there is at present no single greater obstacle than statism to the development of critically personalistic thinking, either as a philosophical framework within which to revive scientific psychology or as a general conceptual orientation for the conduct of everyday social life.[2] What is so problematic here is not the interest in psycho-demographic knowledge *per se*, for such knowledge does have its rightful place both in social science and in community life. The problem is the view that knowledge of prevalent statistical patterns within sub-populations of individuals, which is precisely what psycho-demographic knowledge is, must by its very nature be knowledge that advances our understanding of the individuals who populate those subpopulations, and thus knowledge that does and should inform our understanding of social—i.e., *inter-personal*—matters.

This view is as widely shared as it is utterly mistaken—a decidedly untoward combination—so the point bears re-emphasizing here that demographic knowledge is knowledge of populations *as such*. Period. It cannot validly be understood as knowledge that informs us about *any one* of the individuals comprising those populations.[3] Failures, both among social scientists and within the general public, to understand this basic point verily glare in the ubiquitous custom of treating knowledge of *personkinds*, e.g., males, females, seniors, youths, Blacks, Whites, etc., as if it were, at one and the same time, knowledge of the *persons* who are seen to instantiate those personkinds, i.e., real individuals who happen to be (or, if deceased, to have been) male, female, elderly, youthful, Black, White, etc.

In the previous chapter, this point was developed at some length as it bears on selected aspects of contemporary discussions of race and racial matters in American society. There I explained both that and why presumptions to knowledge about White or Black *persons* based on statistical knowledge of (or beliefs about) White or Black *populations* are both logically invalid and interpersonally prejudicial. Here, I am simply underscoring the fact that such logical invalidities and interpersonal prejudices arise *whenever* the distinction between knowledge of populations and knowledge of individuals within those populations is transgressed.

In any universe where the ideal of truth matters, logical invalidities are epistemically untenable, and thus compromise the integrity of efforts toward that ideal. Likewise, in any universe where the ideal of justice matters, interpersonal prejudices are morally untenable, and thus compromise the integrity of strivings toward that ideal. Compromises of both sorts are widespread today, both in social science and in the public domain. Moreover, as the novelist and social critic Wendell Berry (b. 1934) reminds us, it is important to see that the ideals threatened by these conceptual compromises are not unrelated:

> The pursuit of justice becomes dangerous when it is dissociated from the pursuit of truth ... . Prejudice is opposed to justice, and defies it, by shortcutting or overriding justice's prescribed dependence on truth and the pursuit of truth. (Berry, 2022, p. 44 in Kindle edition)

I submit that the problematic aspects of personkind thinking warrant the sustained and careful critical reflection, not only of psychologists and other social scientists but also of thoughtful and concerned citizens.

Other Troublesome Aspects of Statist Thinking

Unvarnished expressions of prejudices, whether by word or by deed, are not the only obstacles thrown up by statism to the development of a critically personalistic socio-cultural ethos. On the contrary, there are modes of social exchange that appear, at least superficially, to be free of prejudice but are still based on inherently impersonal statistical considerations.

Prevailing Misunderstandings of Probability

One of those ways entails the false belief that knowledge of the proportional frequency of some doing, X, within a representative sample of a given personkind reveals the 'probability' that an individual seen to be of that personkind does (or did, or will do) X. This mistaken notion was discussed in Chapter 5, where its historical roots were seen to lie in a misappropriation of the concept of 'penchant' originally formulated by the early 19th-century scholar Adolphe Quetelet as the basis for his social physics (Porter, 1986). That concept is still widely misappropriated today, though the French term 'penchant' is rarely encountered.

Concern with Typicality

A close relative of false understandings of 'probability' is encountered when attention is focused on the 'typicality' among designated person-kinds of certain ways of doing or being. As noted earlier, 'typicality' of this sort among designated personkinds is a purely demographic notion, and so, in and of itself, has no proper place at all in efforts to advance inter-*personal* understandings.[4] However, when the concept of 'typicality' within some designated personkind is injected into discussions about—and perhaps even directly with—a particular individual of that kind, the implicit presumption is that the population-level phenomenon is relevant to an understanding of something that individual does, or of some way that she/he is. This putative relevance is grounded in the belief that the 'typicality' of some way of doing or being within some designated personkind may be taken as a valid indicator of the strength of the inclination toward that way of doing or being within the psychological makeup of the individual under discussion.[5]

To be sure, it could be sensible from the standpoint of social psychological theory to hypothesize that individuals adopt their particular ways of doing or of being based on knowledge of (or beliefs about) what is typical among their respective kinds of people (or, to borrow a currently popular locution, among people who 'look like them.'[6] However, sound scientific practice would require testing that hypothesis—i.e., exposing it to the risk of empirical disconfirmation—on a case-by-case basis, and this is something that is virtually never done.[7] To simply assume that a population-level typicality will leave its isomorphic trace on the psyches of individual members of that population compromises both the scientific quality of our understandings of those individuals and the fairness of the judgments we make of those individuals as agents in their own development as persons and as fellow citizens. Critically personalistic thinking facilitates the recognition and avoidance of these missteps, and thus enhances prospects for both scientific soundness and social fairness.

Agenting of Personkinds

On occasion, statist thinking issues in exchanges that ought to prompt vigorous head-scratching. A clear if admittedly somewhat banal example of this can be seen in the online weather service application AccuWeather advisory that air measured as 'fair' in quality

> is generally acceptable for most individuals, ... [but] sensitive groups may experience minor to moderate symptoms from long-term exposure.

It is the wording of the latter segment of this advisory that is so puzzling, for it ignores the basic reality that statistically defined categories of people *do not breathe*. As such, those entities are not subject to 'exposure' to atmospheric conditions, nor can they *experience* 'moderate symptoms'—or anything else. What makes the wording in this portion of the advisory all the more curious is its deviation from the wording of the first segment of the advisory, where reference is made, just as it should be, to the experience of 'fair' air quality by (most) *individuals*. Unlike statistically defined categories, individuals *are* entities that breathe, hence *can* be subject to exposure to atmospheric conditions, and *can* experience symptoms. Why the utterly inappropriate and entirely unnecessary shift in wording from the first segment of the advisory to the second?[8]

As acknowledged above, the wording of an air quality advisory is a relatively minor matter. However, in using it as an example here, I hope to heighten appreciation for the conceptual problem that lies at its core, and, in turn, for the societal importance of communicating with one another sensibly about matters of mutual concerns. The ease with which that objective is undermined by statist-inspired locutions is obvious here, resulting in a narrative that may seem sensible on its surface but in fact is not. How we speak with one another both reflects and influences how we think about ourselves and one another, and how we exchange with one another going forward regarding matters of common concern—including those not mundane.

Take, for example, the issue of social justice. That ideal is not something that can be realized in any concrete sense by a category of persons, any more than inferior air quality can affect the respiratory health of a category of persons (Porter, 1995). Social justice—and, of course, its opposite, injustice—can be realized concretely only by persons. Social justice is a moral issue, and our grasp of it is not enhanced by referring to it as if it were a statistical/demographic one. It is not, and this is true even if our concern over the issue has been prompted by familiarity with (or beliefs about) some demographic state of affairs.

Arguably, it could be meaningful to say that social (in)justice prevails (or not) within some category of persons, but *'prevalence'* here would refer to a statistical fact about a collective of persons regarded as a single entity, and not to any flesh and blood person's concrete experience. Where personal doings or states of being are of concern, focus must be scrupulously maintained on persons, not on personkinds, and this is true whether the substantive concern is with the effects of air quality on respiration or with the effects of certain institutional practices or social customs on the experience of (in)justice by the persons living in a community.

There is an instructive parallel to be noted here between the mistake of agenting *aggregates* of persons as if they were persons, and the mistake in the 'other direction,' as it were, of regarding *parts* of persons as persons. This latter mistake has been identified by Bennett and Hacker (2003) as the 'mereological fallacy.'

Mereology is an intellectual discipline devoted to the relationship between parts and wholes, and Bennett (a philosopher, b. 1939) and Hacker (a neuroscientist, b. 1939) have strained to make clear why so much of the contemporary neuroscience literature, both in scientific/technical publications and in the popular press, reflects that error in thinking whereby psychological attributes properly attributed to whole persons are mistakenly attributed to parts of persons, usually brains or brain parts. Appealing directly to the psychophysical unity of persons, a key tenet of critical personalism (refer to Chapter 2), Bennett and Hacker write:

> Neuroscience can … discover the neural preconditions for the possibility of the exercise of distinctively human powers of thought and reasoning, of articulate memory and imagination, of emotion and volition … . What it *cannot* do is *replace* the wide range of ordinary psychological explanations of human activities in terms of reasons, intentions, purposes, goals, values rules and conventions by neurological explanations. And it *cannot* explain how an animal perceives or thinks by reference to the brain's, or some part of the brain's, perceiving and thinking. For it makes no sense to ascribe such psychological attributes to anything less than the animal as a whole. It is the animal that perceives, not parts of its brain, and it is human beings who think and reason, not their brains. The brain and its activities *make it possible* for *us*—not for *it*—to perceive and think, to feel emotions, and to form and pursue projects. (Bennett & Hacker, 2007, pp. 7–8, emphasis in original)

Later on in the same text, Bennett and Hacker (2007) pose challenging questions:

We know what it is for human beings to experience things, to see things, to know or believe things, to make decisions, to interpret equivocal data, to guess and form hypotheses ... . But do we know what it is for *a brain* to see or hear, for *a brain* to have experiences, to know or believe something? Do we have any conception of what it would be for *a brain* to make a decision? (Bennett & Hacker, 2007, p. 18, stress in original)

It is as mistaken to regard aggregates of persons as persons in their own right[9] as it is to regard parts of persons, such as brains or regions of brains, as persons in their own right, and that is why I have thought that this brief consideration of the writings of Bennett and Hacker might help to clarify my argument above.

Empiricist Understandings of the Saying 'Everyone Is Different'

Still another widespread notion in need of careful critical reflection finds expression in the oft-heard saying that 'everyone is different.' On its face, this notion seems fully consonant with the critically personalistic tenet that every person has a distinctive individuality. Yet we have already seen that while statist/personkind thinking can pay lip service to this notion, such thinking will inevitably compromise fidelity to the notion at some point, for it demands that every someone be regarded as *indistinct* from at least some other some-ones, namely those who are judged to be of the target person's same 'kind.' It is only on this basis that personkinds can be defined at all, let alone compared with each other through implicitly or explicitly statistical considerations.

The critically personalistic understanding of distinctive individualities is based not on the *empirical* consideration of extant differences between individuals, but rather on the *conceptual* differentiation between what is judged to be true of a given individual and its hypothetical negation. The characteristic of honesty, for example, can emerge in someone's judgment as distinctive of a given individual against the alternative possibility of dishonesty. On this understanding, actual instances of dishonesty on the part of other individuals need not be—'now' or even, in principle, ever—demonstrably extant within a population.

This is a *rationalist* understanding of the distinctiveness of individualities, and it runs contrary to the *empiricist* understanding that has long been so widely (though uncritically) held among both social scientists and lay persons. According to the empiricist view, the human mind is initially a blank slate, a *tabula rasa*, and all of one's ideas are ultimately the gift of experience. Of course, this would include one's ideas about what might otherwise be the case regarding a particular individual. The following illustration of the rationalist view by the philosopher William Barrett (1913–1992) may help to persuade skeptical readers of the plausibility of that alternative to the empiricist view.

In this passage, Barrett is discussing the revolutionary thinking of the astronomer Galileo Galilei (1564–1642). Barrett explains that in order for Galileo to advance his ideas, he needed a decisive and clear-cut concept of inertia as a fundamental characteristic of moving bodies.

> What does Galileo do? He does not turn to 'irreducible and stubborn' facts [of experience]; rather, he sets up a concept that could never be realized in actual fact.
>
> Imagine, he says, a perfectly smooth and frictionless plane; set a ball rolling upon this plane, and it will roll on to infinity unless another body and force interpose to stop it.
>
> Well, experience never presents us with perfectly frictionless surfaces nor with planes infinite in extension. No matter; these conditions supply us with a concept of inertia more fruitful for theory than any that would be yielded by the 'irreducible and stubborn' facts themselves. Rationalism does not surrender itself here to the brute facts. Rather, it sets itself over the facts in their haphazard sequence; *it takes the audacious step of positing conditions contrary to fact, and it proceeds to measure [i.e., understand] the facts in the light of the contrafactual conditions. Reason becomes legislative of experience.* (Barrett, 1979, pp. 200–201, italics and brackets added)

On the empiricist view of human cognition in general, reason is constrained by experience. On the rationalist view, as Barrett (1979) so vividly explains in his description of Galileo's ideas, reason goes beyond and guides experience. It is this latter view that is embraced in the critically personalistic understanding of the 'distinctiveness' of individualities as I have described it above.[10] Of course, this view is further tied to the basic tenets of critical personalism discussed in Chapter 2. There, it was noted that the framework is one within which persons are conceived as agents, i.e., as active and purposive partici- pants, in their own development. Further, the developmental process itself is understood to unfold in accordance with the values by which the individual chooses to live. The course of one's development as a person is not fully dictated by one's biological endowment (nature) and/or socio-cultural upbringing (nurture). On the contrary, as an agent in the production of one's own self, a person is understood to make choices reflecting the values that she/he does—and does not—embrace, and this power of choice is understood to exist regardless of one's biological nature and socio-cultural nurture. The *possibility of being otherwise*, i.e., of being other than one is (or is judged to be) at any given point in time, is regarded as ever-present. This tenet is a conceptual prerequisite for any viable notion of human freedom and personal responsibility, and, in concert with existentialists on this point, critical personalists would regard the denial of this tenet as 'bad faith' (cf. Bakewell, 2016).

Because the critically personalistic conception of 'distinctive individualities' is grounded in the consideration of one's values, and because it is divorced from the consideration of extant between-person differences, it admits of the possibility of two distinct individualities being *like* each other *without* compromising the distinctiveness of either of them. One's distinctiveness as an individual is defined by the alignment of one's doings and ways of being with certain values, and the dis-alignment of those doings and ways of being with other values. Another's distinctive individuality is defined in the same way. Hence, to the extent that the value systems defining each individuality overlap, the two individualities converge without undermining the integrity of either, regardless of the extent of overlap. This is the understanding of personal identities Stern sought to advance with his concept of *introception,* the process of one person taking up into one's own goal system the values of others and developing into a microcosm (refer to Chapter 2), and it is on this basis that possibility of a critically *inter*-personal—and decidedly non-individualistic—socio-cultural ethos can be envisioned.

A Rationalist Understanding of Societal Distinctiveness

There is a passage in the book by philosopher Susan Neiman mentioned in Chapter 5 (Neiman, 2019) that further illustrates the difference between rationalist and empiricist thinking about distinctiveness, albeit via metaphor and in reference not to individuals but to societies as wholes.

The passage tells of a conversation Neiman had with a German philosopher identified as Bettina Stangneth (b. 1966), who has authored several works having to do with antisemitism and National Socialism in Germany. The conversation pertained to the atrocities of the Holocaust and how they were being addressed in post-World War II Germany. Stangneth was deploring the view of some in contemporary Germany who oppose further displays of contrition and other efforts toward atonement—such efforts, for example, as the building of Holocaust memorials and the placement, in the sidewalks of Berlin, Hamburg, and other German cities, of so-called *Stolpersteine* or 'stumbling stones.'[11] The opponents of such efforts, Stangneth is quoted as saying, 'are not even particularly secret about [their opposition].' Neiman, fully cognizant of similar currents of opposition in the U.S. to educational efforts aimed at informing pupils about the factualities of two centuries of slavery in America, of post-reconstruction Jim Crow practices, and of the enduring legacies of those realities today, interjected: 'Bettina, may I remind you how things are in other countries?' Neiman quotes Stangneth as responding: 'You're asking about my country. *If the soup is too salty, it's too salty. Doesn't matter if you can't get better soup anywhere else.*'[12]

In her distaste for opposition in Germany to continued efforts to acknowledge historical atrocities, Stangneth was entirely indifferent to the possible existence of similar currents of opposition in the U.S. or anywhere else. She judged as 'excessively salty,' so to speak, a current reality in Germany against her conviction that that reality *could*—and in this case *should*—be otherwise in Germany. The 'saltiness' of extant realities elsewhere was irrelevant to her. The issue, Stangneth was insisting, is not an empirical demographic one—inviting, for example, comparisons to determine which society is 'better' at handling such social tensions— but a *conceptual*—and in this instance a *moral*—one.

Stangneth's thinking on this matter is wholly consistent with the critically personalistic understanding of distinctiveness, not only of individualities, but of a society's ethos more generally. The nature and appropriateness of certain social doings and institutional practices are understood in terms of judgments contrasting 'what is' with 'what might otherwise be.' In instances of moral judgments, the consideration of 'what *might* otherwise be' is centered by a conception of 'what *ought* otherwise be.' In no case is the nature or moral standing of a social practice or institution to be judged by whether it appears 'better' or 'worse' than other extant societal practices. If the proverbial soup is too salty, it's too salty. Period.

By stark contrast to the above, an especially egregious example of statist/demographic thinking about an important social issue arose (as I have been told) during a recent meeting convened to address the mass shootings that have occurred on U.S. school campuses nationwide over the past several years, resulting in the deaths of scores of innocent children. During that meeting, the opinion was voiced that widespread concern over such events should be tempered by the statistical fact that other causes of childhood death are currently more prevalent in the U.S. From a critically personalistic perspective, concern—indeed, moral outrage—over school shootings is appropriate because they happen *and ought not to.* The statistical prevalence of other causes of childhood death in the U.S. is absolutely irrelevant.

Respecting the Diversity of Personkinds While Recognizing the Overarching Commonality of Humans

Nothing in what has been said above should be understood to imply that critical personalism is blind to extant personkind differences.

As much as anyone during his time or since, the scholar who founded the subdiscipline named 'differential' psychology at the outset of the 20th century, Stern himself, appreciated the importance of recognizing and systematically taking into practical consideration the facts of personkind differences (cf. Stern, 1900, 1911).[13] Nevertheless, Stern the empirically attentive differential psychologist never surrendered the commitment of Stern the humanistically grounded philosophical psychologist to the notion of a universal humanness that transcends all personkind differences. Critical personalism is the *Weltanschauung* or *worldview* within which Stern sought to articulate those dual commitments, and the present work is offered as an attempt to preserve and contemporize his efforts. However diverse from each other various individuals and peoples might in fact be, they are all persons rather than things, meaning that they are all purposive, value-endowing agents rather than passive and strictly reactive automatons, and they are all potential collaborators in the ideotelic pursuit of the common good rather than mere survival-of-the-fittest adversaries forced into the base quest for dominance.

Going beyond the now widely prevailing embrace of *multiculturalism* as an orientation to the increasing diversity of today's ever more globalized societies, the contemporary social psychologist Fathali Moghaddam has advocated an *omni-cultural* perspective which, while not denying the realities of multicultural diversities nevertheless prioritizes the recognition of human commonalities. The goal of omni-culturalism, Moghaddam (2012) writes,

> ... is a society in which people are knowledgeable about, and give priority to, human commonalities, but also leave some room for the recognition and further development of group distinctiveness. (Moghaddam, 2012, p. 306)

As I hope to have made clear by now, the philosophical worldview of critical personalism is quite in line with the spirit of omni-culturalism as described by Moghaddam (2012), particularly as the former offers a detailed theoretical framework for understanding both general humanness and the development of distinctive individualities. Perhaps the time for just such a worldview has finally arrived.

In line with this aspiration, I conclude with an observation drawn from a podcast that I accessed through the German radio station *Norddeutsche Rundfunk* (NDR; North German Radio) on December 10, 2007. In that podcast, I found an especially poignant, expression of a critically personalistic vision of a socio-cultural ethos, though I have no reason to believe that the featured speaker in the podcast had any familiarity with William Stern's critical personalism.

The segment of the podcast on which I focus here was devoted to an interview that had been conducted with a man named Erdin Kerdoniç, who was born and raised in Bosnia but had lived in Berlin during the civil war in Bosnia-Herzegovina from 1992 to 1995. At the time of the interview, the 32-year-old Kerdoniç had returned to live in Bosnia, and he was reflecting on his life in Bosnia after the end of open hostilities. The interview was given in Bosnian and broadcast on *NDR* in a German translation that I, in turn, translated into the English version given below. In this excerpt, Kerdoniç expressed the view that I have defended in this work as the basis for a critically personalistic socio-cultural ethos:

> During the war, I was thoroughly Bosnian, and had a more or less confrontational attitude against other ethnicities. This was because of what 'my ethnic group' had experienced. Now, since I have returned to Bosnia, I have noticed that the concept of these ethnicities, so, 'I am Croatian!' or 'I am Bosnian!' or 'I am Serbian' is really of no more help in life. It no longer interests me where a person comes from. I look to see what that person has for human values that would be helpful both to that person and to me to move forward in this region we share in common.
>
> The fact cannot be ignored that we are, so to speak, citizens of a land with different regions, with different religious confessions, with different cultural backgrounds, with different ethnic backgrounds, but none of this should ever be primary, [as in] 'Yeah, I'm a Bosnian! I come from Bosnia!' *No!* I am first of all a human being, and then I am a Bosnian.
>
> I think that this process is going to take some time.

With reference to societies well beyond the confines of Bosnia, Kerdoniç's concluding statement is undoubtedly as true today as it was when he made it well over a decade ago. So, it is high time we get started, and the conceit of this volume is that William Stern's critical personalism could be enormously useful in the effort.

Notes

1 The reader may recall from Chapter 2 that it was for strivings of just this sort that Stern invented the term *ideotelic.* One must bear in mind that the stem of his neologism is not *idio-,* referring to something individual, but *ideo-,* referring to *ideals.*

2 In this latter domain, I think it is also arguable that the spirit of individualism that continues to prevail in America is a major obstacle to the development of critically personalistic thinking. The fundamental incompatibility of the two orientations was mentioned in Chapter 2, and a lengthier treatment of the matter can be found in Lamiell (2021, esp. pp. 15–18 and pp. 55–60).

3 This is true, it should be noted, whether those individuals are being conceptualized as persons in the critically personalistic sense of the term or not.

4 The concept of 'typicality' need not always refer to prevalence within a population. On the contrary, it can sensibly refer to the prevalence of some doing or way of being across multiple encounters with a particular individual, and thus to something *typical of that individual.* This consideration applies as well to the concept of 'probability.'

5 As explained in Chapter 5, explicit acknowledgments of 'exceptions to the rule' of typicality rarely suspend beliefs about how typicality informs about individuals, but serve only as advance—and weak—quasi-apologies for the errors that will inevitably be committed by maintaining and acting on those beliefs.

6 I share Neiman's (2023) characterization of this locution as 'rather childlike' (Neiman, 2023, p. 14 in Kindle edition).

7 Thanks to the brilliant work of contemporary psychologist James W. Grice (b. 1964), a general experimental paradigm has been worked out for conducting studies of the required sort in a scientifically sound manner (see, e.g., Grice, 2011, 2014, 2015). A major step toward the successful revival of scientific psychology could be taken by a discipline-wide adoption of Grice's ideas. Unfortunately, that salutary eventuality has not yet come to pass.

8 Similar considerations apply in the face of the oft-heard rhetorical question 'What kind of person would do such a thing?' *Kinds* of persons don't do things, though persons assuredly do.

9 It must be noted here that Stern's worldview recognized as sensible the possibility of regarding groupings of persons such as families and religious congregations as persons in their own right. However, in this connection the groupings in question were conceived as organic units functioning with purpose in accordance with certain value systems, and not as mere aggregates of individuals with no real connection to one another. In one of his later publications (Stern, 1930), Stern acknowledged that he had encountered considerable opposition within the scholarly community to this aspect of his thinking, and he chose not to pursue it further at that time, altogether secure in the knowledge that his concept of person could and should be meaningfully applied to individual human beings.

10 For a detailed development of this notion within the technical domain of psychological testing, see Lamiell (1987). The reader might also find chapter 9 in Lamiell (2003) instructive in this connection.

11 The *Stolpersteine* are small plaques embedded in sidewalk surfaces informing passers-by of the names of the persons who once inhabited property at the respective locations of the stones, and of the fate of those persons at the hands of the Nazis (e.g., the date on which they were arrested and sent off to a concentration camp).

12 This exchange appears on p. 57 of the Kindle edition of Neiman's book. The emphasis here indicated in Stangneth's response to Neiman's comment has been added by me.

13 It is expected that within a year or so of this writing, the entirety of Stern's pathbreaking 1900 book will be published in an English translation completed by the French scholar Serge Nicolas (b. 1962), with some collaboration from myself.

References

Bakewell, S. (2016). *At the existentialist café: Freedom, being, and apricot cocktails.* New York: Other Press.

Barrett, W. (1979). *The illusion of technique.* Garden City, NY: Anchor Press/ Doubleday.

Bennett, M., & Hacker, P. M. S. (2003). *Philosophical foundations of neuroscience.* Oxford, UK: Blackwell.

Bennett, M., & Hacker, P. M. S. (2007). The introduction. In Bennett, M., Dennett, D., Hacker, P., & Searle, J. (Eds.), *Neuroscience & philosophy* (pp. 3–13). New York: Columbia University Press.

Berry, W. (2022). *The need to be whole: Patriotism and the history of prejudice.* Shoemaker and Company Publishing.

Gantt, E. E., & Williams, R. N. (Eds.) (2018). *The nature and consequences of overreachOn hijacking science: Exploring the nature and consequences of overreach in psychology.* New York: Academic Press.

Grice, J. W. (2011). *Observation oriented modeliing: Analysis of cause in the behavioral sciences.* New York: Academic Press.

Grice, J. W. (2014). Observation oriented modeling. *Comprehensive Psychology*, *3*, ISSN 2165-2228.

Grice, J. W. (2015). From means and variances to persons and patterns. *Frontiers in Psychology, 6*, 1–12.

Lamiell, J. T. (1987). *The psychology of personality: An epistemological inquiry.* New York: Columbia University Press.

Lamiell, J. T. (2003). *Beyond individual and group differences: Human individuality, scientific psychology, and William Stern's critical personalism.* Thousand Oaks, CA: Sage Publications.

Lamiell, J. T. (2018). On scientism in psychology: Some observations of historical relevance. In Gantt, E. E., & Williams, R. N. (Eds.), *On hijacking science: Exploring the nature and consequences of overreach in psychology* (pp. 27–41). New York: Academic Press.

Lamiell, J. T. (2019). *Psychology's misuse of statistics and persistent dismissal of its critics.* London, UK: Palgrave-Macmillan.

Lamiell, J. T. (2021). *Uncovering critical personalism: Readings from William Stern's contributions to scientific psychology.* London, UK: Palgrave-Macmillan.

Machado, A., & Silva, F. J. (2007). Toward a richer view of the scientific method: The role of conceptual analysis. *American Psychologist*, *62*, 671–681. 10.1037/ 0003-066X.62.7.671.

Moghaddam, F. (2012). The omnicultural imperative. *Culture & Psychology*, *18*, 304–330.

Neiman, S. (2019). *Learning from the Germans: Race and the memory of evil.* New York: Farrar, Strauss, and Giroux.

Neiman, S. (2023). *Left is not woke.* Hoboken, NJ: Polity Press.

Porter, T. M. (1986). *The rise of statistical thinking: 1820–1900.* Princeton, NJ: Princeton University Press.

Porter, T. M. (1995). *Trust in numbers: The pursuit of objectivity in science and public life.* Princeton, NJ: Princeton University Press.

Stern, W. (1900). *Über Psychologie der individuellen Differenzen (Ideen zu einer "differentiellen Psychologie* [On the psychology of individual differences (Toward a "differential psychology")]. Leipzig: Barth.

Stern, W. (1911). *Die Differentielle Psychologie in ihren methodischen Grundlagen* [Methodological foundations of differential psychology]. Leipzig: Barth.

Stern, W. (1917). Die Psychologie und der Personalismus (Psychology and personalism). *Zeitschrift für Psychologie und Physiiologie der Sinne,* 1. Abteilung. Zeitschrift für Psychologie, *78,* S. 1–54.

Stern, W. (1924). *Person und Sache: System der philosophischen Weltanschauung, dritter Band: Wertphilosophie* [Person and thing: Systerm of critical personalism, Volume 3: Philosophy of value.] Leipzig: Barth.

Stern, W. (1930). *Studien zur Personwissenschaft. Erster Teil: Personalistik als Wissenschaft* [Studies in the science of persons. Part One: Personalistics as science]. Leipzig: Barth.

Stern, W. (1938). *General psychology from a personalistic standpoint* (H. D. Spoerl, Trans.). New York: Macmillan.

Stern, W. (2010). Psychology and personalism (J. T. Lamiell, Trans.). *New Ideas in Psychology, 28,* 110–134. 10.1016/j.newideapsych.2009.02.005.

Watson, J. B. (1913). Psychology as the behaviorist views it. *Psychological Review, 20,* 158–177. 10.1037/h0074428

Watson, J. B. (1928). *The ways of behaviorism.* New York: Harper and Brothers.

Wundt, W. (2013). Psychology's struggle for existence (J. T. Lamiell, Trans., originally published in 1913). *History of Psychology, 16,* 197–211. 10.1037/a0032319

Index

Note: Page numbers followed by "n" indicate a note on the corresponding page.

'abstract idealism' 90
Achenwald, Gottfried 48n2
allen gemein (*common to all*) 8, 13
Allen, R. T. xvii
Allport, G. W. 19–20, 33, 68
Anastasi, Anne 11
'appropriation' process 29

Bakan, David 38–40
Baldwin, James 81
Barrett, William 98
Behavioral Research: A Conceptual Approach 39
behaviorism 91
Bell Curve, The 87n5
Bennett, M. 96
Berry, Wendell 72
Brightman, Edgar A. 86
Buckle, Henry Thomas 76
Bühring, G. xvi
Burgos, J. M. xvii, 20
Burrow, R., Jr. xvii, xxin3

Caste: The Origins of Our Discontents 80–81
cause-effect relationships 11
child psychology, socio-cultural voice in the domain of 55–62
choice reaction time (CRT) 7
Cohn, Jonas 23, 33, 58, 60
colorblindness in contemporary social exchanges 85–87
Concept of Mind, The 30n2
'concrete idealism' 90
convergence 29

critically personalistic thinking, conceptual obstacles to 90–102; concern with typicality 94; confession of a concrete idealist 90; contemporary statism 92–93; early 20th-Century behaviorism 91–100; 'everyone is different', empiricist understandings of 97–99; personkinds, agenting of 95–97; personkinds, diversity 100–102; probability, misunderstandings of 94; societal distinctiveness, rationalist understanding of 99–100; statist thinking, troublesome aspects of 93–99; then and now 91–100
critical personalism 15, 19–30; *see also* person
CRT *see* choice reaction time

Descartes, René 23
Deutsch, W. xvi, 68n1
DeWolf, L. Harold 86
DiAngelo, Robin 73, 77, 80
'differential' psychology 9–13, 87n4; establishment of 9–11; experimental psychology's transformation into 11–13; methodological foundations of 10; 'nature' and 'nurture' combination 13
discrimination reaction time (DRT) 7

discrimination time (DT) 7
Donders, F. C. 7
Drobisch, M. W. 6, 8, 12, 76, 83
DRT *see* discrimination reaction time
DT *see* discrimination time
Durbeck, P. 31n5, 44–45

Ebbinghaus, H. 16n3, 20–22, 24, 48n5
empirical psychology 3–15; early
    experimental psychology,
    illustration of 7–8; original
    structure of 6–8; to psycho-
    demography transformation
    3–15; psychological doings of
    individuals 14–15;
    see also differential psychology
    species; *psycho*-demography
Epstein, S. 27
experimental psychology 11–13

Fechner, Gustav 16n3
Feger, B. 59
Foss, M. A. 31n5
Freeman, Mark 46
Freud, S. 59

Galilei, Galileo 98
Gergen, Kenneth J. 16n2
Glaude, Jr., Eddie S. 81
Grice, James W. 103n7

Hacker, P. M. S. 96
Hanson, F. A. 66
Harré, Rom 81
Hegel, G. W. F. 90
Heinemann, R. xvi
Hempel, A. 31n5
heterotelic (hetero-telic or others'
    goals) 28
History of Civilization in England, A 76
Human Personality, The xvi, 20, 27
Husserl, Edmund xvii, xxin2

ideotelic 102n1
impersonal aspects of discourse about
    race in America 80–87
individualism, personalism versus 20
inter-*personal* relationships 3–4
Interpretation of Dreams, The 59
introception 99
Introduction to Personalism, An xvii
Introzeption (int(e) roception) 29

Kerlinger, F. N. 39–40, 44
King, Martin Luther, Jr. 85
Koczanowicz-Dehnel, I. xvi
Krenshaw, K. W. 82–84, 87n7

Lamiell, J. T. 31n5, 44–45, 103n10
Larsen, R. J. 31n5
*Learning from the Germans: Race and
    the Memory of Evil* 84
Leffel, G. M. 31n5
Löwisch, D.-J. xvi
Lück, H. E. xvi

Martin, J. 46
mental chronometry 7–8
Meumann, Ernst 58
Moghaddam, F. 101
multiculturalism 101
Münsterberg, H. 9, 67

'naive personalism' 23
Neiman, S. 84–85, 103n6
'nomothetic' knowledge 40–41
Nunnally, J. C. 48n12

Observation-Oriented Modeling
    (OOM) 44
'own-goals' *autotelic* ('auto-telic')
    28

Pavlov, Ivan 91
person 20–30; as a causally effective
    agent 24–26; concept of 20–30;
    definition 22–23; as a
    distinctive individuality 26–28;
    heterotelic (hetero-telic or
    others' goals) 28; hypertelic
    goals 29; ideotelic goals 29; as
    an inherently evaluative being
    26; 'own-goals' *autotelic*
    ('auto-telic') 28; *own* goals
    (Selbstzwecke) 28; parts 22;
    personal being, critically
    personalistic perspective
    22–23; person–thing
    distinction 21, 23; person
    world convergence 29;
    psychosocial development
    from a critically personalistic
    perspective 28–30;
    rudimentary considerations
    20–30; syntelic goals 28;

teleological perspective 25; as
a unitary, psychophysically
neutral being 23–24
personal identity, impersonal
understandings of 82–85
Personalism: A Critical Introduction
xvii
personalism, individualism versus 20
Person and Thing 26, 90
personkinds xx 93; agenting of 95–97;
diversity of 100–102; personal
nature of discourse about
71–80
Porter, Theodore M. 48n2, 73–74
probability concept 73, 103n4;
prevailing misunderstandings
of 94
psychoanalytic psychotherapy with
children and adolescents 59–62
psycho-demography 4–5, 16n5;
'differential' psychology
establishment 9–11; empirical
psychology's transformation
into 3–15; psychology's
transformation into 9–13
psychological doings of individuals
14–15; population-level
statistical regularity 14; versus
individual differences in
psychological doings 14–15
psychological inquiry 3–15; need to
revive 3–15
psychological studies, reviving,
challenge of 33–47; confusion
within a confusion 40–42;
expanding space for
qualitative investigations
45–47; 'hybrid' discipline 47;
'nomothetic' knowledge
40–41; from psychological
science to scientistic
'psychology' 35–42;
restructuring psychological
experimentation 43–45;
reviving psychological science
42–47; 'science' as 'the making
of knowledge' conception,
reviving 47
psychological testing 62–66; socio-
cultural concerns about the
proliferation of 65–66; as a
socio-cultural issue 62–65;

socio-cultural voice in the
domain of 62–66
*On the Psychology of Individual
Differences: Toward a
'Differential' Psychology*
9, 62
*Psychology's Misuse of Statistics and
Persistent Dismissal of its
Critics* xix
Psychology's Struggle for Existence 35
psychosocial development: from a
critically personalistic
perspective 28–30; *own* goals
(Selbstzwecke) 28
'psychotechnics' 48n1

Quetelet, Adolph 6, 73

racism in American Society xviii–xix,
71–87; colorblindness in
contemporary social
exchanges 85–87; critically
personalistic observations on
71–87; impersonal aspects of
contemporary discourse about
80–87; mechanistic narratives
concerning 80–82; personal
identity, impersonal
understandings of 82–85;
personkinds, impersonal
nature of discourse about
71–80; statism in White
fragility 77–80
*Recollection, Testimony, and Lying in
Early Childhood* 56
Rousseau, Jean Jacques 56–57
Ryle, Gilbert 30n2

Schiff, B. 46
scientism 36
simple reaction time (SRT) 7
social dynamics of systemic racism,
mechanistic narratives
concerning 80–82
societal distinctiveness, rationalist
understanding of 99–100
socio-cultural voice, William Stern's
53–68; in the domain of child
psychology 55–62; highly
talented pupils, identification
58–59; lying in children,
development 56–58;
psychoanalytic psychotherapy

with children and adolescents
59–62; in psychological testing
domain 62–66; tolerance,
ethical significance of 53–55
SRT *see* simple reaction time
statism 92; contemporary statism
92–93; in White fragility 77–80
statisticism 34
statist thinking, troublesome aspects of
93–99
Stein, Edith xvii
Stern, W. xvi, xviii–xix, xxin2, 9–11,
14–15, 16n6, 19–22, 24–26, 33,
36–38, 46, 53–68
syntelic goals 28

tachistoscope 7
Taylor, F. W. 68n4
teleophobia 25
Thorndike, E. L. 10
Titchener, E. B. 16n3

tolerance, ethical significance of 53–55
Trierrweiler, Foss 31n5
Twain, Mark 34
Tyler, Leona 11
typicality 94, 103n4

*Uncovering Critical Personalism:
Readings from William Stern's
Contributions to Scientific
Psychology* xix

Venn, John 75

Watson, John B. 36, 91
White fragility, statism in 77–80
*White Fragility: Why It's So Hard for
White People to Talk About
Racism* 73, 77
Wilkerson, I. 80
Windelband, Wilhelm 40–42
Wundt, W. 6, 35–38, 48n5, 91
Wundt-ian style experimentation
11–12